CABLE:

An Advertiser's Guide to the New Electronic Media

Ronald B. Kaatz

CRAIN BOOKS   740 Rush Street   Chicago, IL 60611

Published by Crain Books
A Division of Crain Communications, Inc.
740 Rush Street
Chicago, IL 60611

84 83 82 10 9 8 7 6 5 4 3 2 1

ISBN: 0-87251-076-X

Library of Congress Number: 82-072511

Printed in the United States of America

With love to Suzi, Kathy, and Roberta

Contents

Figures **viii**

Tables **x**

Preface **xii**

1 The Early Days **1**

A Century of Developments: the 1880s to 1980s **2**
The Origins of Cable **2**
Early Rules and Regulations **4**
Who Owned the Systems: Investors and Finance **4**
The Early Days of Cable Advertising **5**
More Rules and Regulations **7**
Advertising Moves Several (Small) Steps Ahead **9**
The Breakthrough: Satellites, Pay Services,
   and the Superstation **12**

2 Cable: Today and Tomorrow **17**

Cable and Pay Cable **17**
STV, MDS, and SMATV **18**
Pay-Per-View **20**
Multichannel MDS **20**
Satellites **21**
Direct Broadcast Satellites **22**
Low-Power Television **22**

Video Cassettes and Video Discs   **23**
Video Games   **24**
Personal Computers   **24**
Interactivity: Qube and Videotex   **25**
Bigger Screens, Better Pictures, Sensational Sound   **27**
A Multitude of Choices   **28**

3   The 1980s and Beyond: New Options for All   **29**

Diversity of Choice   **29**
Diversity of Time   **31**
Interacting with the TV Set   **33**
An Issue of Support: Who Pays?   **37**

4   The Impact of the "New" Media
on the "Old" Media   **39**

Television   **39**
Radio   **43**
Newspapers   **44**
Magazines   **45**
Out-of-Home   **48**
Media and the Future   **48**

5   The New Media Planning Process   **49**

What Can Cable Do for You?   **50**
Understanding Cable's Deficiencies   **52**
Why Get Involved Now?   **54**
Establishing Cable Advertising Objectives   **57**
Implementing a Cable Advertising Program   **59**
Getting the Job Done   **60**
Achieving Cable's Maximum Benefits:
   Case Histories   **62**

6   Different Avenues for Different Advertisers   **71**

The Scope of the Business: National or Local   **71**
The Nature of the Product or Service:
   Mass-Market or Non-Mass-Market   **77**
Two Approaches to Cable:
   Media Audience and Response   **82**
Cable and the Direct Marketer   **87**
Information in the Home   **90**
Your Imagination: The Only Limitation   **97**

7 Creating and Producing Advertising
for the New Media **101**

Concentrating on Message Content **102**
The New Media Message That Motivates **105**
Understanding and Creating the Infomercial **110**
Tapes, Discs, and the Future **114**

8 Measuring Results: Qualitative and
Quantitative Approaches **115**

The Need for *Facts* . . . Not Just *Faith* **115**
A Question of Technique: What to Measure **118**
A First Step—Counting Cable Households **121**
Some Basic Problems Facing Cable Measurement **123**
The Creative Cable Research Kit **128**
Cumulative Audience: A Helpful Tool **135**
Be a Cable Research Explorer **139**

9 Cable at the Local Level **141**

The Local Impact of Cable **141**
What's Happening Out There? **142**
The Selling of Local Cable Advertising **145**
The Local Commercial Production
Problem—One Solution **147**
Local Cable Advertising and
Program Production Guidelines **149**
Conclusion **154**

Appendix 1: The Local Cable Idea Starter Kit **155**

Appendix 2: Satellite Network Buying Checklist **165**

Appendix 3: Cable Satellite Networks **169**

A Glossary of 121 Essential
New Media Advertising Terms **187**

Sources of Additonal Information **199**

Index **201**

Figures

2–1. Possible Entertainment Distribution Pattern of the Future **21**

3–1. New Options for Television **31**

4–1. Commercial Flipping and Zapping **42**

4–2. Unchanged Viewing Patterns **42**

4–3. New Viewing Patterns **43**

4–4. New Viewing Patterns **43**

5–1. Cable Video—One of Many Advertising Options **51**

5–2. Opportunities Offered by Cable **52**

5–3. Developing Strategy for Selecting and Using the New Media **57**

5–4. Making a Good Cable Buy **59**

5–5. Setting and Implementing Cable Advertising Strategy: Step One **60**

5–6. Setting and Implementing Cable Advertising Strategy: Step Two **61**

5–7. Setting and Implementing Cable Advertising Strategy: Step Three **61**

5–8. Setting and Implementing Cable Advertising Strategy: Step Four **62**

5–9. Setting and Implementing Cable Advertising Strategy **63**

6-1. Media Communications Segments **79**
6-2. A Late-Fringe Television Buy vs. A Dog Show **88**
7-1. TV and Cable Communication **103**
7-2. The Cable Paradox **104**
7-3. Producing the New Media Message That Motivates **106**
7-4. Traditional Broadcast Production Schedule **107**
7-5. The Infomercial **111**
7-6. Infomercials: Long and Short **113**
8-1. Correct Base for Cable Audience Measurement **126**
8-2. Kraft Cable Questionnaire **134**
8-3. Cable System Subscriber Questionnaire **136**
8-4. Advertising Impact Scoreboard **138**

Tables

1–1.   Cable Penetration, 1969   **7**
1–2.   Advertiser Interest in Cable, 1970   **11**
1–3.   Up on the Satellites   **14**
2–1.   The Growth of Cable   **19**
2–2.   A Forecast of New Media Coverage   **27**
3–1.   Stations/Channels per Home per Year   **30**
3–2.   Number of Channels Received   **30**
3–3.   A Possible New Media Viewing Schedule   **32**
5–1.   Cable Penetration in the Top 20
        U.S. TV Markets (Percent)   **53**
5–2.   A Checklist of "New Media" Media Values   **58**
5–3.   Who Buys Cable?   **59**
5–4.   Achieving Cable's Maximum Benefits   **65**
6–1.   Cable Penetration in the Top 20
        U.S. TV Markets (Percent)   **72**
6–2.   Advertiser-Supported Satellite Services   **73**
6–3.   Cable Markets of Opportunity   **74**
6–4.   Nielsen Coverage Areas with over 50 Percent
        Cable Penetration (Percent)   **77**
6–5.   Readership of Upscale Magazines by Potential CBS Cable
        Households Compared to National Average   **80**

6–6.	A 50-Channel Cable Prototype	**83**
6–7.	A 10-Year Forecast of New Media Impact on Network Audiences (Percent)	**84**
6–8.	1981 Prime-Time Network Ratings in Cable and Noncable Homes (Percent)	**84**
6–9.	1981 Cable and Noncable Home Network Ratings for Different Dayparts	**85**
6–10.	Possible Ratings for Cable and Noncable Homes (Percent)	**85**
6–11.	Past 4-Week Reach and Frequency	**86**
6–12.	Present 4-Week Reach and Frequency	**87**
6–13.	U.S. Videotex Tests and Services	**92**
6–14.	Possible Vehicles for Different Products and Services	**94**
6–15.	Sales Volume of Leading Direct Marketers, 1981 (estimated figures)	**95**
8–1.	Who Views Cable, Pay Cable, and STV	**116**
8–2.	CNN Distribution by Nielsen Territory	**122**
8–3.	CNN Compared to All Television Households	**122**
8–4.	A Cable Research Comparograph	**125**
8–5.	Multiple Channels to Measure	**127**
8–6.	Characteristics of "Daytime" Female Cable Subscribers Compared with All Adult Women	**129**
8–7.	Reading Preferences of Audience in CBS Cable Area	**130**
8–8.	Results of "Dial-It" Telephone Poll	**133**
8–9.	Arbitron Weekly Cumulative Adult Ratings	**138**
8–10.	VideoProbeIndex Cumulative Adult Ratings	**139**
8–11.	WTBS Cumulative Household Ratings	**140**
8–12.	Four-Week Reach for One Cable Satellite Service (Assuming One Commercial per Program)	**140**
9–1.	Local Selling of Spots on Advertising-Supported Networks	**147**

Preface

If someone had suggested in 1972 that a book be written on advertising and video technology, the immediate response would have been, "But what will you say after the first couple of pages?"

And yet today, more is written by more reporters in more periodicals about Videotech™ (or what is more generally called the New Electronic Media) than has been written about any other advertising or marketing subject in decades. An entire new lexicon of terminology has been created—concepts and words that either didn't exist or were discussed only in the most technical circles a few years back. A dozen or more new magazines have sprung up to deliver Videotech™ news and features to the communications industry, advertisers, and viewers. And older publications have created special new media sections and departments. Hardly a week goes by without a meeting, seminar, conference, or convention devoted to cable television or home video. These events all underlie the growing importance of the new video technology to everyone involved in the dissemination of information and the marketing of goods and services.

In the years following World War II, no two developments more significantly shaped the world of advertising than television and the computer. With television, marketers at last had available to them a medium that could bring their products to life and present them instantly to millions of consumers. And, later, the computer gave these same marketers the ability to process and analyze millions of pieces of information on the media habits, demographic characteristics, and purchasing patterns of these prospective customers.

The "Third Wave" of progress is now upon us. It promises to have as great an impact on everyone involved in the production, distribution, and sale of goods and services as have television and the computer.

Television provided advertisers with the means of reaching millions of people at a single point in time in an entertainment environment. The new video technology allows these same advertisers to zero in more closely on thousands of their best customers in an environment that is more finely tuned to their product or service. At the same time, the new media opens up the world of video to advertisers whose narrow marketing targets, confined geographic sales areas or limited budgets precluded them from television.

Actually, the new electronic media represent more than a single medium. They encompass an entire spectrum of means to deliver entertainment, information, and advertising into our homes via the television set. These include cable television, video information services, and home video players and recorders.

From an advertiser's standpoint, the arrival of these new media selections creates not only new opportunities, but new problems as well.

Cable: An Advertiser's Guide to the New Electronic Media focuses mainly on cable television, the most developed of the new media from the standpoint of advertising potential. At the same time, it offers suggestions as to how the other new media services can also be used to communicate advertising and information in an effective manner. And, of course, as it focuses on the opportunities, it will also help you solve some of the problems you will encounter as you begin to travel down the new media advertising road.

In a nutshell, *Cable: An Advertiser's Guide to the New Electronic Media* is targeted at everyone involved in planning, buying, selling, and developing advertising for—or just learning about—cable and other new media. It is an "idea starter" and is aimed at getting you to start thinking about how to use the new media to communicate in new and creative, yet practical and down-to-earth ways.

As you read it, keep an open mind and jot down your own thoughts and ideas that can be applied to buying or selling your products or services.

Look at this book as a cross between a roadmap and a cookbook. It will show you the routes available to get from here to there and offer "recipes and information about the preparation of nourishing and tasty (New Media) food."

I would like to close on a special note of thanks. Writing a book involves the cooperation of many people—probably more than I could name. But two people deserve special mention: Richard Hagle, my Crain Books editor, helped me develop the concept of this book and keep on track in its execution; and Marilyn Foster not only could read my nearly illegible handwriting but also knew what I meant to say when my writing was *totally* illegible.

Thank you both!

CHAPTER 1

The Early Days

The new media of the 1980s encompass a veritable gourmet menu of viewing options and video gadgetry. There are actually so many options that many homeowners are converting what once was a family room or den into a home video center.

At the June 1982 Consumer Electronics Show in Chicago, Sony and Zenith, among other major consumer electronics marketers, started to push component TV systems that are much like component audio systems. *Advertising Age* reported:

> Modular design makes it possible to mix and match video components to build a video system that suits individual tastes and needs.
>
> With it, the long-promised home entertainment center of the future is on its way to reality. The promise envisions a big, sharp, 45-inch or more projection TV screen as the center of activity, with other components such as a video cassette recorder, video disc player, video games, home computer, and stereo. (June 7, 1982, p. 3)

Everything would be controlled from a single source selector allowing for the instantaneous delivery of entertainment and information complete with stereo sound.

A Century of Developments: the 1880s to 1980s

It may surprise you, but as we focus on advertising and the new media of the 1980s, we really go all the way back to the 1880s and the earliest beginnings of television. It was then that a patent was taken out on a rotating disc that had spiral perforations through which it could scan a scene. Unfortunately, nothing was really done with this device until half a century later. Then, during the 1920s and 1930s, an Englishman named John Baird developed a television set that people could buy in London department stores. Baird broadcast programs using an improved version of the old rotating disc and even figured out a way to record his images on what had to be the world's first "video disc." Unfortunately, neither the discs nor the "Televisor" set itself bore much resemblance to today's TV sets. More importantly, no one was either ready or very much interested in developing advertising for them.

The real beginnings of television, as we know it, probably took place in San Francisco, when Philo Farnsworth first demonstrated *color* television in 1927. A key role was then played by Vladimir Zworykin, who invented the first *electronic* TV camera pickup tube during the 1930s. This brought television out of the labs and into the spotlight where it was demonstrated to mass audiences at New York's 1939 World's Fair.

World War II held up the production of sets and the growth of stations, but once the fighting ended, the new medium took off. Sets were coming off the assembly lines, and stations were growing throughout the country. All that was missing were programs to fill the time. Very quickly, however, the networks came along and not only did this, but, in the process set in motion the creation of the most powerful advertising device that had yet been imagined — the television commercial.

The Origins of Cable

Why CATV Came About: Reception and Variety

Television had barely gotten underway in the late 1940s when the first new media seeds began to sprout. It was back in 1948 that the earliest cable systems were born in remote areas of Pennsylvania and Oregon. Known then as Community Antenna Television (CATV), its function was simply to bring TV signals into communities where off-air reception was either nonexistent or very bad because of interfering mountains or distance. The earliest systems were often started by appliance dealers so they would be able to sell television sets to the folks in their community.

In Astoria, Oregon (5,000 homes), it is reported that Leroy Edward Parsons wired the first system with an initial subscriber list of three households. The first commercial CATV system was installed two years later in Lansford, Pennsylvania, after which they began springing up all over the map. To bring in the signals, CATV systems employed master antennas that were mounted on tall towers erected on the plains or on mountain tops. The cable system operator then delivered the signals to subscribing homes in the community via coaxial cable that was either strung overhead on utility poles or underground in conduits.

In small communities without any television service, CATV provided a brand new medium. And, in many other towns that had only one or two television stations, viewers now had the full program offerings of three national networks. Cable even helped reception in the cities—in San Diego, which is encircled by mountains, and in Manhattan with its giant steel buildings. The term *subscriber* was aptly applied to those homes that were hooked up to cable. They paid an installation charge to be connected, and then they paid a monthly fee of $3 to $5 just as if they were subscribing to a local newspaper.

In the 1950s, the system operator's master antenna was joined by microwave relay equipment that extended the range of originating station pickup. Common carrier terrestrial microwave routes emanated from major market areas and transmitted the network signals to cable systems hundreds of miles away. Early on, it was recognized that a function of CATV could be to supply channel diversification. Suppliers like Jerrold Corporation developed sophisticated equipment that permitted reception of 6 to 12 channels.

While these advances were well received by viewers, many television station operators (network affiliates) viewed diversification with alarm. In the early 1960s, programming was "bicycled" to many smaller market stations by film (and later tape). If a cable system could import a network program weeks in advance of a local network affiliate's own scheduled airdate, the local station's own audience delivery could be affected seriously. And, of course, the emergence of the independent television station in the larger metropolitan markets—

and its subsequent microwaving to the small communities—further impacted on a local station's ratings and, hence, what could be charged for advertising time.

Early Rules and Regulations

The concerned voices of station owners were heard very loud and clear in Washington, and in 1964 the Federal Communications Commission issued its first cable regulation. Cable operators were required to "black out" programming that came in from distant markets and duplicated a local market station's own programming, if the local station demanded it. Just as a point of reference, there were only about one million cable homes in the entire country at the time this took place! (It is interesting to note that even in the early days of cable, "audience fragmentation" was a factor that stations had to worry about, even though they didn't know it by that term.)

Despite the imposition of FCC regulations, cable systems continued to grow throughout the country. They served a much-wanted function of bringing program diversity to areas of limited program availability. The broadcasting industry, however, was still alarmed by the potential of cable, and in 1968 the FCC, for all practical purposes, imposed a "freeze" on its further growth. The new ruling said that if a cable system was going to be located within 35 miles of a television market, it could offer its subscribers only those TV signals that were available off air (i.e., not broadcast). This served to bring cable development in the larger markets to a virtual standstill. To a certain extent, cable was once again relegated to fringe areas and markets with poor off-air reception. As another benchmark, there were fewer than three million cable homes at this time. This represented about 5 percent of the television homes in the country.

Who Owned the Systems: Investors and Finance

In its earliest days, CATV systems were generally owned singly by local businesspeople. They were small—in 1952 there were 70 CATV systems with a total of 14,000 subscribers (or 200 subscribers per system). It was not unusual for a local hardware store or electrical contractor to own a system. After all, he had the materials, equipment and know-how to build it. This image of CATV as a "small business community service" was soon altered as large groups quickly realized the profit potential in the industry. Because of popular demand, liberal depreciation policies, and debt-leveraged ownership, many early systems showed rates of return on invested capital exceeding 15 percent, while returns as high as 50 percent were claimed by a few. And this was before the days of the added pay services! It is no wonder that cable was quickly discovered by many of the largest names in the communications industry. In contrast to its early "Ma and Pa" status, the cable industry today is dominated by a number of major companies known as MSO's (Multisystem Operators). The Top 15 cable MSO's today account for over half of the industry's total subscribers.

The Early Days of Cable Advertising

With only 5 percent television home penetration, CATV systems still had substantial influence on the broadcasting industry as the 1960s came to an end:

- Most apprehension of CATV expressed by the networks was oriented to the future. Market fragmentation and possible audience loss and, hence, financial injury to their local affiliates began to concern them. There was even concern that a "fourth network" of CATV systems might be formed to compete in the advertising marketplace.
- To many broadcasters, the growth of CATV posed a large threat to the survival of many local stations, especially the UHF stations, which were the most vulnerable to competition. In a 1968 decision, the FCC ruled in favor of the local stations and restricted the expansion of CATV in San Diego. (Even so, Cox Cable in San Diego is the largest system in the country today.)
- The subject of copyright was rearing its head as CATV operators took broadcast programs off the air without permission of the music and program copyright holders. At the same time program costs to broadcast stations whose audiences were being boosted by cable were increased.

Cable's impact on advertising in the late 1960s was not yet very significant, but many marketers were already looking ahead to the problems and opportunities that would be created:

- Advertisers foresaw possible audience losses to their spot buys that would result from the viewing of distant stations by CATV subscribers in the market.

- On the other hand, advertisers might gain "bonus" coverage through transmission of a station's signal from one market to another. Unfortunately, such coverage might not be desirable. For example, a McDonald's or Burger King promotion in Denver might be carried by CATV into western Nebraska where the item was not available.

- CATV could create a new, inexpensive local advertising medium if cable operators sold commercial time to help pay for cable-originated programming.

- While CATV had little effect on most advertising agency media buys, any future reshuffling of markets and market fragmentation would seriously confuse the time-buying function.

- The problems of audience measurement were not yet major, but it was recognized that they would become significant for the rating services as CATV spread. There were already certain smaller markets (Salisbury, Santa Barbara, Clarksburg-Weston, Marquette, and Missoula) where CATV penetration was over 40 percent and where stations were beginning to feel their audiences were not accuartely reflected in the ratings.

Probably the greatest use of cable by advertisers in these early days was for advertising testing. It started early in 1964 when the Center for Research in Marketing in Peekskill, New York, contracted with a Port Jervis, New York, CATV firm to use its commercial testing facilities. The project was possible because of a "split cable." The basic concept of the test system was fairly simple. Test commercials were slipped into regular network programs as the shows were fed via cable into subscriber homes. The subscribers were split into two separate but demographically matched samples. The researchers varied copy, media weight, commercial length, frequency, etc., and studied the effect of all this on the two samples of viewers. Unfortunately, because the supermarket scanning devices now in use in such tests had

TABLE 1–1. Cable Penetration, 1969

Nielsen Market Areas	Percent CATV Penetration of TV Homes
Major Metro Areas	1.6
Medium Sized Cities	4.8
Smaller Towns	16.8
Rural, Farm Areas	10.2
Total United States	6.4

Source: A. C. Nielsen, November 1969.

not yet been invented, the precision of these tests was limited.

By the end of the early days of cable, the industry had 2,260 systems and 3.6 million subscribers. Overall, 1 out of 16 homes in the country was cable equipped, with the majority of these homes located in smaller towns and rural communities. (See Table 1–1.)

More Rules and Regulations

As the 1960s drew to a close, the big question concerning CATV was, "What might exist tomorrow?" While the outlook for CATV seemed favorable, its future was clouded by just what direction FCC regulatory proposals would take. Three possible scenarios for the future existed:

- If a series of restrictive laws or decisions were produced, CATV could be expected to do no more than serve those areas of marginal reception it was originally created to serve.

- If regulatory decisions were favorable *and* if there was a real demand for a broader choice of programming and not just clearer reception, CATV would expand into the major urban markets.

- Finally, if favorable regulatory decisions and market demand were joined by increased expenditures by subscribers and system operators, the services offered by CATV might expand to include data retrieval, video newspapers and magazines, banking, and home shopping.

In 1972, the FCC set in motion a gradual deregulation of the cable industry to permit the growth of its subscribers and programming. Cable systems had to carry all local market station signals. But, in addition:

- Operators in the Top 50 markets could carry up to three network and three independent stations.
- In markets 51-100, the rule was three network and two independent signals.
- In smaller markets, systems could bring in three network signals and one independent signal.

The new regulations also included what was known as the "leapfrogging" rule. Very simply, it said that the stations brought into a market had to come from one or both of the Top 25 markets nearest to the cable system. An operator could not "leapfrog" over one market to pick a preferred station from another market. If this had been possible, a "Superstation" could have been born several years ahead of Ted Turner's WTBS in Atlanta.

While CATV operators had the green light to carry local signals and import distant ones, existing local market television stations (within 35 miles of the CATV operation) would receive considerable protection in the form of program exclusivity. For example, a CATV system could not duplicate local market programming via an imported signal from a distant station. It would have to "black out" such programming if it was also carried by a local station in the market. In terms of sports, the FCC also ruled that cable systems could not carry games that were blacked out on local network stations normally carried on the system. And, finally, cable systems in the Top 100 markets would be required to spend money to have a minimum capacity of 20 channels, to have built-in two-way capacity, and to provide the public and local government with at least one channel gratis for educational and communications purposes.

Even with this so-called "lifting of the freeze," the implementation of cable's expansion was still impeded somewhat by two factors. First was the considerable backlog of applications before the FCC. Second was the copyright controversy, for while cable operators had agreed in principle to copyright payments for the programs they carried, they and the copyright owners had yet to agree upon a specific system of fees.

Advertising Moves Several (Small) Steps Ahead

In this climate of potential future growth mixed with still-unresolved issues, 5,000 members of the cable industry met in Chicago in May 1972 for the 21st Annual Convention of the National Cable Television Association. There were now six million subscribers, and in an atmosphere of "happy confusion," the industry began to tackle the many problems that would need to be solved if cable was to become a significant communications medium and not just a small-town mover of signals. One such topic was advertising.

There was little to indicate that cable operators were really interested in the national advertising dollar in 1972. This was most evident in the absence of any significant number of program suppliers. The emphasis was largely on barter, movies, free films, or low-budget talk and interview projects. The most ambitious program offering was a Julia Meade series that offered advertising and would be available on cable systems with three million subscribers. (Remember that this was several years before satellite delivery of cable programming.)

There was discussion of CATV's potential on a PI (per inquiry) basis as electronic direct mail. For the most part, however, the real future of the medium as seen in 1972 was felt to lie in the area of strong local-interest shows. Advertising opportunities could develop here, and advertising agencies could take advantage of using their local offices, field representatives, and regional broadcast buyers as information sources.

Considerable attention was focused on pay television, with two systems announcing its introduction in San Diego, Sarasota, and Vancouver. The FCC wanted to encourage pay TV, but the real key to its success was a supply of product — namely big, new movies. Even at this still relatively early point in the expansion of cable, there was discussion about the limited use of commercials on pay television to divide shows into three acts and partially subsidize its costs, reducing prices to subscribers.

The relative lack of interest that cable operators displayed in advertising during the early 1970s was really not surprising. After all, CATV, unlike broadcast television, had a revenue source all its own — subscriber revenue. More important, the system operators were mechanically and not marketing oriented. They could string cable, but very few knew how to sell commercials, much less produce them.

Before cable could successfully attract and hold advertising revenue, it would have to document its ability to compete with other television opportunities in the efficient delivery of an advertiser's prime prospects. And cable operators would have to be shown how advertising could become a significant source of revenue and how they could develop it without either jeopardizing their subscriber revenue or putting in an undue amount of effort.

To put cable advertising at this time into perspective, in 1970 only 500 out of 2,500 systems were capable of originating programs; another 1,500 could provide automated originations such as time, weather, news, and stock ticker reports. Advertising was carried on only 57 of the cable systems that originated programming and on approximately 400 more that provided automated services. The average commercial cost about $15 a minute, and advertisers paid under $100 for an hour show. The best estimate of total cable ad revenue for 1970 was $3 million, a figure that compared with $300 million in subscriber revenue and well over $3 billion spent for broadcast television advertising.

Advertisers' use of cable followed a variety of paths:

- General Foods, Campbell Soup, Hudson Vitamin, and RCA Records tested cable in five small markets via "Monitel." Owned 49 percent by Readers' Digest, Monitel offered a 24-hour feed of time, temperature, weather, general information, and national advertising. The test was inconclusive, and Monitel ceased operating.

- In New York, sports events, including the New York Knicks and Rangers games, were cablecast from Madison Square Garden on Sterling Manhattan Cable and Teleprompter. Avis, Schaefer Brewing, Miles, L&M Cigarettes, and Warner-Lambert each spent about $100 a spot for the coverage.

- In Pittsfield, Massachusetts, McDonald's sponsored the cable system's coverage of the annual Fourth of July Parade.

- On Lower Bucks Cablevision in Levittown, Pennsylvania, ads appeared 72 times each day at 30-minute intervals on the automated "TeleShop'r" service. The cost of spots in this feed of horoscopes, menus, and related incidental information was about 6 cents each.

- An early "how-to" show was a series of 13 instructional programs — "Salt Water Game Fishing." Teleprompter scheduled it three times a week on its Manhattan system and sold advertising time.

TABLE 1-2. Advertiser Interest in Cable, 1970

	National Advertiser	Local Advertiser
Automated time, weather, news	No	Yes
Older syndicated shows already run locally	No	Yes
Cable shows aimed at a small, select audience (How-To)	Yes	Yes
Local community programming	Some	Yes
Programs produced by larger MSO's (a while off)	Yes	Yes

Source: A. C. Nielsen, 1970.

These represented a few of the early ways in which advertisers were using cable. Table 1–2 indicates advertiser interest.

For the present, however, viable opportunities that would attract the national or regional marketer were extremely limited, and cable did not yet offer potential advertisers an efficient means of advertising. Furthermore, even when advertisers did purchase time in CATV programming to take advantage of the medium's selectivity, audience delivery was very difficult to evaluate.

Viewed from the early 1970s, the potential of cable was largely in the future. But, as a J. Walter Thompson Company report noted in January 1973:

It is not too soon to be thinking of future uses which might enable advertisers to capitalize on the unique opportunities which cable will offer as a direct response medium:

- With two-way capabilities, immediate response to advertising is possible. The vehicle might be a "how-to" show tied to a particular product or service.

- The cable family can be identified and reached by at least one monthly mailing from the operator. Advertising material could be included within the mailing.

- Cable could serve as a classified medium.

- Cable could provide a research tool for measuring advertising communications effectiveness.

Considerable discussion was taking place about two-way cable, the growth of cable networks, and the development of pay television via cable, and its audience selectivity was on the minds of many. In September 1970, Benton & Bowles' top media executive George Simko said:

> There is the opportunity to develop and place programming that is geared to very specific audiences defined by job function, lifestyle, etc. On this basis, it is conceivable that television advertisers could isolate highly selective marketing targets in much the same way as selective magazines now perform this function. However, here again the broader programming availability could also serve to fragment the mass television viewing audience still further, leading to greater out-of-pocket expense required to deliver current levels of audience. . . .
>
> If they could produce highly specialized programming, programming of interest to doctors, let's say, advertisers might be willing to pay more than they do for other media and measure the cost on another yardstick. (*Marketing/Communications* [April 1971], p. 31.)

By the mid-1970s, approximately 3,500 cable systems were serving about ten million subscribers. Unfortunately, however, the tremendous costs of wiring the major markets and the fact that most people there got relatively good over-the-air reception was holding back the growth of the industry. In addition, system owners who hoped to reap large revenues with Pay Television couldn't sell Home Box Office to someone who hadn't signed up for the basic cable service. Many large cable companies suddenly faced the prospect of losing vast amounts of money.

The Breakthrough: Satellites, Pay Services, and the Superstation

A breakthrough was needed — something that would transform CATV and its improved signal carriage into *cable television* with a vast new menu of viewing opportunities. This breakthrough occurred in 1975-76.

The first significant event took place on December 12, 1975, when the Satcom I communications satellite was launched. This provided a highly cost-efficient means of distributing multiple program options to cable systems across the country. Programming could be beamed from the ground 22,300 miles up to the satellite and then down to receiving earth stations (or "dishes") at cable systems across the country.

By providing instant reach of nationwide cable audiences, the satellite made it financially attractive to offer programming that would attract subscribers. It was HBO that first capitalized on this when in late 1975 it took a gamble and began distributing the first pay television mix of movies, sports, and specials via satellite to cable systems nationwide.

In those areas not yet cabled, pay television services were developed with distribution over the air. In some markets this was known as Subscription Television (STV). A scrambled television signal was sent

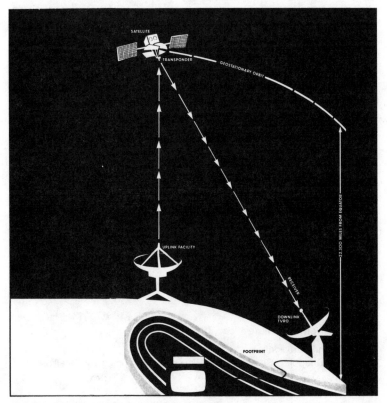

A basic satellite communications system

TABLE 1–3. Up on the Satellites

The number and variety of programming services available to cable systems has exploded since HBO had its satellite debut in 1975. Viewers have a wide array of options from the subscriber supported services as well as from the advertiser supported networks.

Premium Entertainment and Movies
Bravo, Cinemax, Disney Channel*, Entertainment Channel, EROS, Escapade/Playboy, Galavision, Home Box Office, HTN Plus, Movie Channel, Private Screenings, Public Subscriber Network (PBS)*, Showtime, Spotlight

Superstations
WTBS (Atlanta), WGN (Chicago), WOR (New York)

Sports
ESPN, USA Sportstime

News
Cable News Network, Cable News Network 2, C-SPAN, Dow Jones Cable News, North American Newstime, Satellite NewsChannels, Weather Channel

Ethnic and Special Interest
Black Entertainment Television, English Channel (USA), Spanish International Network, Telefrance (SPN)

Culture and the Performing Arts
ARTS, CBS Cable

Education and Public Service
Appalachian Community Service Network

Women, Family, Health and Service
Cable Health Network, Daytime, USA Daytime

General Entertainment, Information and Multi-Features
CBN Satellite Network, Modern Satellite Network, Nashville Network*, Satellite Program Network, USA Network

Religion
Eternal Word Television Network, National Christian Network, National Jewish Television, People That Love, Trinity Broadcasting Network

Music
Apollo Entertainment Network*, Heartbeat Media Network*, MTV: Music Television

Children's
KidVid Network*, Nickelodeon, USA Kidstime

Shopping and Participation
Home Shopping Show (MSN), PlayCable, Shopping by Satellite*, The Shopping Game*, UTV Cable Network*

*Scheduled for launch.

---

over the air and unscrambled by a decoder box on the viewer's television set. In other areas, Multipoint Distribution Systems (MDS) transmitted a pay service via microwave for generally short distances to subscribers who picked them up with special antennas and converter boxes. The interesting point is that although the long-term potential of "nonbroadcast television" was felt to be in its delivery of a wide variety of very special services and programming, it was the movies, with their widespread mass appeal, that really got things moving.

While all of this was going on, Ted Turner, an imaginative, fast-talking businessman who owned the Atlanta Braves, and a 24-hour independent Atlanta television station, came up with the idea of using a satellite to distribute his station's programming to cable systems across the country. The only thing standing in his way was the FCC's leapfrogging rules. This would prevent a cable system from taking his station if there was an independent in a Top 25 market nearer to him that could be carried. But in January 1976, the FCC eliminated the leapfrogging rules, and in December of that year, the nation's first Superstation, WTCG (now WTBS), was up on the bird and flying!

Thus, in only about a year, more steps had been taken to provide cable with the resources and impetus to leap ahead than had occurred in the first 25 years of its existence.

During the last third of the 1970s, earth receiving stations rapidly popped up at cable systems across the country. Their growth was spurred on by sharp reductions in their cost and by a rapid acceleration in the number and variety of satellite services. This, in turn, sped the household growth of existing cable systems and increased the pace of new cable construction throughout the country. (See Table 1–3.)

Cable was ready to enter the 1980s!

CHAPTER 2

The New Media: Today and Tomorrow

Television of the 1980s is a far cry from that of earlier days. Sets are available that fill a wall or fit into a pocket. They can be plugged into a car's cigarette lighter or operated by a battery. They even have telephones built in so you can watch and talk at the same time. In the 1980s, what has been known simply as "television" is becoming "viewer-controlled video." In the process, viewers will find themselves no longer prisoners of what is on and when it is on. Rather, they will watch what they want when they want it with dozens of program choices and a variety of video recording and playback devices.

Let's briefly examine what elements make up the new video mix.

Cable and Pay Cable

Cable television today reaches one-third of the homes in the country. Of all the new media forms, it is today the most dominant in terms of viewer penetration and advertising potential. Until recently, cable was largely a small-town, at most a 12-channel, proposition used to bring TV to rural areas and improve reception. This is rapidly changing as the cable companies franchise the big U.S. cities and construct 36-, 52-, and even 104-channel systems.

17

Cable subscribers pay a monthly fee of about $8.00 to supplement their basic network entertainment diet with a wide variety of special programs. For approximately an additional $9.00 a month, subscribers may purchase what is referred to as a "pay cable" service. This consists mainly of movies with some sports and special programs —uncut, uncensored, and (at least for now) uninterrupted by commercials. Many cable systems today offer "tiers" of more than one pay service, and it is not unusual to find movie buffs subscribing to two or more of them. Their homes in essence become "multiscreen cinemas."

Subscribers to pay cable services have long been regarded as representing very lucrative marketing targets. They are younger, better educated, and substantially more affluent than the population at large. As a result, considerable interest has been expressed by advertisers and advertising agencies in having commercial opportunities available on pay cable.

It is debatable whether or not the major *existing* pay services will become advertising carriers, even if the commercials are uninterruptive (between rather than within films) and of a unique, entertaining nature. The pay services realize that one of their attractions to many subscribers is their noncommercial nature. If only 1 percent or 2 percent of their subscribers cancelled because of the commercial intrusion on them, the services and their systems could lose more in revenue than they might ever expect to recoup in advertising income. (See Table 2–1.)

New pay cable services, however, may well begin operations with a limited amount of advertising and offer subscribers a lower monthly cost than *existing* services. In addition, advertisers will probably find numerous opportunities to reach pay subscriber homes by advertising in pay cable program guides. It will be a case of using a print medium to reach a video audience.

STV, MDS and SMATV

In large cities where cable television has not yet arrived, pay services are often available via Subscription Television (STV) and Multipoint Distribution Systems (MDS). These are single-channel services for which subscribers pay about $20 per month plus an installation fee.

Signals are delivered over the air via one of two techniques:

- Subscription Television involves delivering a scrambled picture from a UHF station that is unscrambled in subscribers'

TABLE 2–1. The Growth of Cable

| | Basic Cable | | | Pay Cable | |
Year	Cable Systems	Subscribers (millions)	Percent of TV Homes	Subscribers (millions)	Percent of TV Homes
1952	70	.014	>1	—	—
1960	640	.550	1	—	—
1965	1,325	1.275	2	—	—
1970	2,490	4.500	8	—	—
1975	3,506	9.800	14	—	—
1976	3,681	12.100	17	.600	1
1977	3,832	13.200	18	1.500	2
1978	3,875	14.200	19	3.000	4
1979	4,150	16.000	21	5.300	7
1980	4,225	18.700	24	7.800	10
1981	4,400	22.600	28	11.800	14
1982	4,800	27.400	33	14.800	18
1985	5,500	36.000	42	23.000	26
1990	6,500	54.000	58	37.000	40

Source: Extrapolated from *Television Digest, Cablevision.*

homes via a decoding box. This is the technique that ON TV offers over Channel 44 in Chicago.

- A Multipoint Distribution System also employs an over-the-air system, but rather than coming from a UHF station, it is from a closed-circuit microwave transmitter. Again, a scrambled signal is unscrambled by the decoding box on top of the set, but this time it is played through an empty channel. In Chicago, Showtime is an MDS system available to dwellings within 10 miles of the Hancock building.

STV and MDS systems are available today in about 2 percent of the nation's households. Penetration will probably peak at about 5 percent and then remain fairly stable. Unlike cable, which brings a large multichannel menu of programming into a home, STV and MDS systems represent a single channel with movies as the major attraction. Still, for people who see cable in their city only as a distant light at the end of a long franchising tunnel, STV and MDS systems offer a viewing option *today*.

For the most part, STV and MDS homes can be expected to convert to cable as soon as it is available to them. However, enough homes will remain unwired to allow 5 percent penetration to hold at least through the last half of the decade. Assuming 90 million television households by that time with a monthly subscription rate at the current $20 level, over-the-air pay television would still represent a $1 billion annual business, even with this relatively low level of penetration.

In a number of major markets where cable may not yet be available, still another service has been developed—Satellite-Fed Master Antenna TV. SMATV is a minicable system for buildings connected to a private satellite antenna. It provides multichannel video to large apartment buildings and condominium complexes.

Pay-Per-View

There are strong indications that an area of high potential in terms of future movie and special event exposure is what is referred to as "pay-per-view." Since homes must be specifically "addressed" and fed individual programs for which they are charged, it requires special equipment.

It is just beginning, but by the mid-1980s, television viewers may be able to pay to see one-time showings of films that have just completed their initial runs in theaters. The distribution pattern that develops may be: movies would be released to theaters first, then broadcast on a pay-per-view basis, followed by home distribution in video cassette and disc form. Next would come pay cable and STV and MDS. Only after these stages would some of the films come to free broadcast television (obviously with significantly reduced audience potential).

Multichannel MDS

In an attempt to challenge cable and provide more than a single entertainment channel, Microband Corporation, the nation's largest multipoint distribution service common carrier, has applied to the FCC to establish multichannel MDS services in the Top 50 markets. If approved, it will be "an urban over-the-air wireless cable network" and will, besides the usual movies and sports, offer two-way information and data services plus pay-per-view movies. Three such operations, each providing five channels of service, will be sought in each city, and fees charged subscribers would be about the same as cable with a one-time charge for a special antenna.

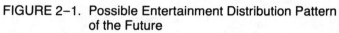

FIGURE 2-1. Possible Entertainment Distribution Pattern
            of the Future

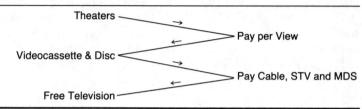

A multichannel MDS system would provide a new service in noncabled areas *and* competition to cable operators in wired areas. It would also be an additional "viewer competitive force" with which local broadcast stations would have to contend.

Satellites

On September 30, 1975, Home Box Office transmitted to its subscribers the Ali-Frazier heavyweight fight live from the Philippines. The unusual thing about this transmission was that it was via Western Union Westar satellite (or "bird") locked in geostationery orbit, 22,300 miles over the equator. The impact of communications satellites on the cable industry has been dramatic. In 1982, some 40 program services are being transmitted to receiving earth stations at cable systems across the country by RCA's Satcom, Western Union's Westar, and AT&T's Comstar satellites. And in 1983, these will be joined by a new Hughes Communications satellite—Galaxy I—devoted entirely to the transmission of cable programming.

Satellite transmission and its diversity of program offerings have had a major impact in spurring ahead cable franchising in the major markets. And, as the cost of earth stations has come down so that "do-it-yourselfers" can assemble one for under $5,000, it is not unusual to see an occasional backyard "dish" picking up programming directly off of a satellite for the benefit of its owner. It's the latest status symbol!

Satellites are also very big business. In late 1981, with a quarter-scale model hanging where Renoirs and Rembrandts are generally displayed, Sotheby Parke Bernet auctioned off seven-year leases for space on RCA's Satcom IV satellite that was launched January 12,

1982. At prices ranging from $10.7 million to $14.4 million a slot, the auction took only 10 minutes and brought in $90 million. It was a classic case of the demand for satellite space far exceeding the supply. (Later, however, the FCC invalidated the auction. They said it would result in price discrimination with different buyers paying different amounts for the same service—and that RCA would have to offer all of Satcom IV's transponders at the same price.)

Direct Broadcast Satellites

By mid-decade, the very newest of the new media forms should be underway. Direct Broadcast Satellites (DBS) will send a signal from a satellite directly into individual homes via small two-or-three-foot rooftop earth stations. With a decoder, an individual will be able to pick up three or four channels that will feature all sorts of specialized programming, such as movies, sports, the performing arts, children's shows, and special interest programming. DBS will offer more variety than single-channel STV and MDS but substantially fewer viewing options than cable. Unlike cable however, the enormous investment of stringing cable will be avoided. Still, the cost of setting up a DBS system is not inexpensive. It is estimated that it will take as much as $1 billion for a system to get into operation!

Low-Power Television

While the stakes involved in getting into cable or DBS are enormous, a plan by the FCC may shortly enable an individual to establish a television station for well under $100,000. The technology behind what is known as low-power television is not new. For many years, low-power transmitters have amplified and rebroadcast the weak signals of distant major-market stations to viewers miles away in rural areas. Now, the FCC is permitting the use of this technology to allow the building of low-power, or mini, television stations with broadcast ranges limited to 10 to 20 miles. The FCC hopes this will open up the commercial airwaves to women, minorities, and other new voices.

When the FCC began accepting applications for low-power stations, they received over 5,000, including one from a young Chicago Black whose plans involved a program lineup aimed at the city's Blacks, Chinese, Mexican, Asian, and Polish population. Applications also came in from religious groups, unions, and a successful business woman. At the same time, though, some large companies were represented. Allstate Insurance Company (Sears) owns 50 percent of Arizona-based Neighborhood TV Company, which hopes to distribute country-and-western programming via satellite to 141 low-power stations across the country.

From an advertising standpoint, marketers should watch the development of low-power television with great interest. It can provide excellent opportunities for targeting very specific messages to some very specific audiences. In a sense, it could develop much like local radio with thousands of stations nationwide.

Video Cassettes and Video Discs

Up to this point this discussion has covered only new media programming that is transmitted, in one way or another, to a viewer's television. Video cassettes and video discs are two forms of new media that do not involve direct transmission to an individual's television.

The video cassette recorder (VCR) plays prerecorded material (largely movies that can be bought or rented) and, when used with a tape camera, can take home movies. The newest video cassette format uses a quarter-inch tape, versus the conventional half-inch. It is fully portable and combines a camera and cassette recorder in one lightweight unit. It can also tape off air.

Unlike the VCR, the video disc player cannot record. Like a phonograph, which the discs resemble, it can only play back. The more expensive laser models have stereo and special features such as freeze-frame, frame-by-frame advance, slow motion, fast motion, and fast scanning. The less expensive stylus models do not have all of these features. And, of course, as is the case with video cassette recorders, the recordings on one system cannot be played on another "incompatible" system.

Since the video disc player cannot record, what would be its advantage over the VCR? *First*, it is the cost of the discs relative to tapes. A movie tape might cost up to $80 while its cost on disc would be under $25. Since most people rent tapes at $2 to $5 a day, this may become academic. The *second* possible advantage is the outstanding picture quality and, in the laser models, the stereo sound.

On the optical (laser) disc, there are 54,000 still frames that can be stored on *each side*. With the ability to access each of these frames individually, there is an enormous number of ways in which the disc might be used in the dissemination of information. The entire works of the great artists of all time, encyclopaedias, dictionaries, travel guides—all these and much more can be stored on very few discs. Their marketing potential was recognized by Sears when it developed a video disc version of its Summer 1981 Catalogue that allowed customers to study stills of the products as well as moving-action sequences, and, of course, all of the necessary prices and specifications.

Video Games

Today's video games have come a long way since the introduction, a few years ago, of Odyssey's Pong games with the bouncing dots. Programmable units allow a person to choose from dozens of games with incredibly realistic representations of space ships, playing fields, and games and players, as well as great control over strategy. It is not even all fun and games any more, as manufacturers are now offering educational cartridges.

Video games have also found their way onto cable. PlayCable, derived from Mattel Electronics' IntelliVision™ Intelligent Television, is an all game channel that offers its subscribers 24-hour, 7-day a week access to a choice of 15 different video games at a time for a fraction of their retail price in stores. On a regular basis, new games are added to an assortment of continuing and returning games drawn from five Intellivision program networks: Sports, Action, Strategy, Gaming, and Learning.

While many would find it difficult to imagine advertising opportunities on video games, the potential is there. With their interest in sports, a soft drink company, in conjunction with Atari or Mattel's IntelliVision™, could develop a new video football game with built-in video "sponsor" identification. The advertiser's involvement would hold down the game's cost to the consumer, and would also be available as a premium device.

Personal Computers

Personal computers have been developed largely for small businesses headquartered at home and professionals such as doctors, lawyers, and dentists. Today, individuals use them to organize their financial records, do their taxes, control energy, and do dozens of other projects. While little has been done in terms of developing advertising potential, the potential undoubtedly is there. In particular, "how-to" programs with product tie-ins might be offered on an almost limitless variety of subjects.

H&R Block could offer a program for maintaining home financial records. At the end of the year, it would be delivered to the nearest H&R Block office to be used by H&R Block in preparing an individual or family's tax return. No more last-minute gathering together of records on scraps of paper!

Interactivity: Qube and Videotex

The last third of the 1980s will be known as the Era of Interactivity. As new wiring is laid and smaller 12-channel cable systems are upgraded to 36, 52, and 104 channels, two-way cable will expand.

The most publicized and the first interactive cable service is Qube. Qube was developed in Columbus, Ohio, by Warner Cable and has expanded to Cincinnati, Pittsburgh, Dallas, and all other markets in which Warner-Amex has been awarded franchises. A marriage between the television set and the computer, Qube enables the viewer to respond to messages on the screen. They can answer questions asked on the screen, take part in instant polls, and even purchase products from their living rooms.

Another category of interactive services comes under the generic classification of Videotex. It includes two services—Teletext and Viewdata.

In Teletext, alphanumerics and graphics are carried over the unused picture linage of a television channel and delivered page-by-page as selected by the viewer. On a regular TV channel (such as KNXT in Los Angeles, one market where it was tested), Teletext has a capacity of only 200 pages. When it is given its own full cable channel, the capacity grows to about 5,000 pages.

On the other hand, with Viewdata, the alphanumerics and the graphics are stored in a computer and delivered over telephone lines or cable. The viewer can select this material and interact with it. The potential number of pages of data that can be generated is limited only by the capacity of the computers employed.

Viewtron completed a test of a Viewdata system in 1981 in Coral Gables, Florida, and is expanding the test to 5,000 Miami homes in 1983. The tests are underwritten by Knight-Ridder with technology supplied by AT&T. The system is simple to operate and allows subscribers to select what they want through the conventional "tree-branching" process, a constant narrowing down from broad categories of information to more and more precise data. The user has a key pad that is used like a telephone to communicate with the system and request information. Or, requests can be made by using a typewriter-like keyboard to request, say, W-E-A-T-H-E-R or D-O-W I-N-D-E-X.

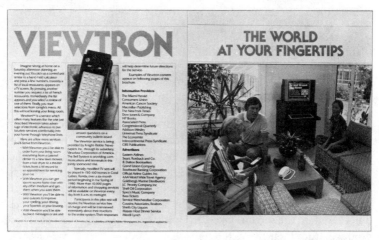

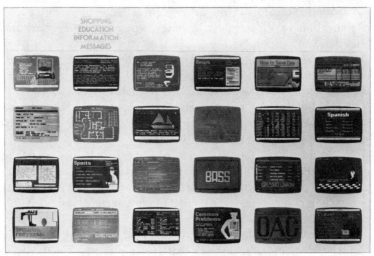

Viewdata tested its Viewtron system during the summer of 1982 in Coral Gables, Florida.

Both Teletext and Viewdata have considerable advertising potential and can be used to supply information on products and services. Viewdata can be used for direct home shopping in which the home customer can select what he or she wants to purchase from a variety of options and actually place an order. In the Viewtron test, for example, Sears, Ward's, and Penney's were among the experiment's participants.

Bigger Screens, Better Pictures, Sensational Sound

Just as changes are taking place in what can be done on the television screen, so are there significant innovations in the hardware itself. Matsushita Electric, Japan's largest consumer electronics company, has been developing:

- A TV screen 8½ x 11 feet that can be used for either front or rear projection.
- High Definition Satellite TV system for receiving super-high-frequency (SHF) signals transmitted simultaneously to the entire country. With 1,125 scanning lines rather than the 525 currently in use in the U.S., the picture quality is far sharper than what we now see.
- Television with stereo and bilingual sound.
- Three-dimension video cassettes in which the viewer wears stereoscopic glasses to see the effect. An airline might use this to develop an entire series of 3-D travel films for promotion purposes.

TABLE 2–2. A Forecast of New Media Coverage

	Percent of All TV Homes	
	1982	1990
Cable	33	58
Pay Cable	18	40
STV and MDS	2	5
DBS	—	5
Text Information Systems	—	10
Video Cassette Recorders and Video Disc Players	4	20
Video Games	18	30
Home Computers	1	20
Giant Screens	>1	10

A Multitude of Choices

In a nutshell, the 1980s will be known as the decade of "a multitude of media choices." New video hardware and a variety of video software will combine to expand both the number of options available to the viewer and the flexibility with which he or she can exercise these options. And by the time the decade ends, we will be looking at the widespread blooming of many new media that only recently have begun to sprout. (See Table 2–2.)

CHAPTER 3

The 1980s and Beyond: New Options for All

Advertisers today must become knowledgeable about the new electronic media from the standpoints of diversity of viewing choice (new options for using the TV set), diversity of viewing time (time-shifting potential), and interactivity.

More than anything else, the 1980s will offer advertising prospects a new *diversity* of video experiences. With diversity of *choice* of both information and entertainment, consumers will no longer be restricted to a handful of offerings available from the networks and independent stations. And diversity of *time* means they will no longer need to alter their lifestyles to fit network or station schedules. Viewers will watch what they want to see when they want to see it *or* are available to see it.

Diversity of Choice

Viewer choice has steadily increased over the past 30 years as a result of the growth in the number of broadcast stations and cable systems. In 1981, the average home could receive two and one-half times the number of stations it did in 1953 (see Table 3–1). Looked at another way, over half the homes in the country could receive 10 or more channels in 1981 while one quarter of the homes received 13 or more channels. (See Table 3–2.)

And this is only the beginning of choice, since two out of every three cable homes in 1981 were hooked up to systems with 20 or fewer channels. New systems now under construction will offer 36, 52, or 100+ channels. And the older 12-channel systems will eventually be upgraded to provide the same larger variety of selections.

Add to the growing number of cable channels the anticipated infusion of low-power television stations coupled with the arrival of direct broadcast satellite transmission. It is not out of line to forecast that by the end of the decade, viewers will be able to choose from a vast array of video offerings in the same manner as they now select

TABLE 3–1. Stations/Channels per Home per Year

Year	Average Number
1953	3.8
1960	5.7
1970	6.8
1981	9.2

Source: A. C. Nielsen, 1981.

TABLE 3–2. Number of Channels Received

Number of Channels	Homes Receiving (Percent)
1–4	8
5	8
6	9
7	8
8	6
9	8
10–12	29 ⎱ 53
13+	24 ⎰
	100

Source: A. C. Nielsen, 1981.

what to listen to from a multitude of AM and FM radio stations.

Of course, for those desiring even greater choice, video cassettes and discs offer additional alternatives just as stereo records and tapes provide audio diversification.

Diversity of Time

In Quincy, Massachusetts, a local ordinance forbade dancing on Sunday. Hence, the town's "California Disco" owner decided to show five straight hours of "General Hospital," which he had taped earlier in the week. Working women—whose jobs kept them away from their favorite soap opera Monday through Friday—packed the place. They spent $10 to $12 each on drinks, about the same as did the Saturday night disco crowd. Business was in fact so good that the "California" expanded to two showings a Sunday!

This 1980 "happening" is a classic example of new media "time shifting." And while a relatively small number of homes (under 5 percent) have video cassette recorders or video disc players in 1982, those who do are finding that they greatly alter their viewing habits.

The VCR gives its owner a "timeshift" capacity by which he or she can record a program at one time and play it back at another. Days and hours blur together as viewers watch at their leisure and on their own time schedules rather than on the schedules of the media. And cable satellite networks include multiple exposure patterns—with programs and features repeated within the same day, week, month, or over the course of a year.

FIGURE 3–1. New Options for Television

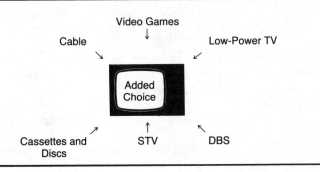

In homes with video recorders, viewing patterns are both altered and shifted as owners buy pre-recorded programming and tape other programs to watch at times other than those of the original broadcasts. One important point is that, as the Nielsen Home Video Index reported, VCR households are upscale in education, occupation, and income. Equally important for advertisers is that Nielsen and other researchers have found that people would accept advertising in video cassettes if the prices of the cassettes could be held down.

Table 3–3 gives an example of how a working woman might create her own video viewing schedule. Over breakfast, she watches Johnny Carson's monologue from "The Tonight Show" of the previous evening. After dinner, she scans her tape of "The CBS Evening News," watching only those items which interest her—fast-forwarding past everything else. Later, she settles in with Phil Donahue, after which she curls up in bed with "General Hospital." With her VCR she has totally time shifted the network and station schedules to meet that of her own lifestyle.

Television viewing may actually develop a bit like magazine reading. People usually do not sit down and read an issue cover to cover. They pick it up at first, look at whatever interests them, and then perhaps return to it later and look at other material. With more viewing alternatives, these same people will be able to look at what they want to see and record other programs to look at in whole or part later.

By the end of the 1980s the claim that television reaches people at one single point in time might no longer be the case for all shows. Just as a magazine accumulates its audience over several weeks (or months), viewers will record certain shows and their audience will

TABLE 3–3. A Possible New Media Viewing Schedule

	Conventional Television	Time Shifted Video
Morning	"Phil Donahue"	"Tonight"
Afternoon	"General Hospital"	—
Early Evening	"CBS Evening News"	—
Prime-time	"Whatever is On"	"CBS Evening News" "Phil Donahue"
Late Night	"Tonight Show"	"General Hospital"

accumulate over time rather than in a single instant. This would have its greatest impact on advertising seeking to reach its audience quickly at one point in time, for example, in the case of special promotions.

Interacting with the TV Set

As viewers gain a new diversity of choice and of time, they will also change the manner in which they interact with the television set. The typically stereotyped *passive* viewer who only watched (sometimes inattentively) will become an atypically *active* consumer of a vast menu of video offerings.

Through two-way interactive cable and other data transmission systems, information will be delivered to the viewer on request and responded to by him. Such capabilities are with us right now.

Home Shopping

Despite its seemingly utopian quality, home video shopping is on its way into our living rooms. Inflation, transportation costs, and crime are all increasing the potential of this new video service. In addition, more husband-wife working households, coupled with the rise of the single adult household, are increasing the need for greater convenience and preservation of one's leisure hours.

There are obviously some products that the consumer will always want to see in person and examine in detail, but there are others which he or she would just as soon buy from home. It is possible that home shopping could drastically alter the retailing structure for a wide variety of these goods and services. For example, an aisle in the supermarket of the future might be a video screen in the home. For perhaps the first time ever, a person can tell at a glance what everything costs. And, of course, unit pricing will also be included. In addition, on every frame will be the name of the retailer in whose "Video Supermart" one is shopping.

The "Video Shopper" is an example of a home shopping service available today. It operates without two-way cable in a relatively simple manner. Consumers receive the "Consumers Video Shopper Catalog." In addition to the information provided in the catalog, "Video Shopper" infomercials are shown on a local cable channel. These infomercials provide additional information on the product and explain special features. To make a purchase, the customer calls a toll-free number and places the order. In Spring 1982, the "Video Shopper" was being carried into homes across the country on Atlanta's Superstation–WTBS.

Home Security

An example of what many people feel will be a highly demanded cable service in the future is home security. Using interactive equipment, it will monitor subscriber homes for smoke, fire, or intrusion. There will also be a "panic button" that will be available in the case of a medical emergency. The entire system will be linked to appropriate agencies and medical facilities.

Another special service is the Emergency Warning Device. When the community-wide alert system is activated, the device sounds in the home. In effect, it tells subscribers to turn on their television sets and tune to a special channel for emergency information.

While these services do not offer on-cable advertising potential, they do provide opportunities for tie-in print promotion. Companies such as Bristol-Myers, Abbott Laboratories, or Johnson & Johnson are ideal candidates for such advertising. Companies in these product categories could prepare home safety and first-aid pamphlets for local cable systems to include in their monthly statements to subscribers.

Interactivity: What the Future Holds

The impact of interactivity will be most felt during the last third of this decade when the hardware needed for it is widely available to homes across the country. In the meantime, a number of interesting experiments are taking place to test its potential.

One such experiment, The CABLESHOP, allows for viewer interaction, with the telephone serving as the request device rather than a special cable converter. J. Walter Thompson USA and Adams-Russell Company, a large telecommunications firm and multisystem operator, tested this interactive information system in 1982 on Adams-Russell's 52-channel cable system in Peabody, Massachusetts. Sixteen national and four local advertisers participated in the test.

The CABLESHOP programmed three- to seven-minute segments with helpful service-oriented advertising and community information. Viewers wishing to participate in the service had a "menu" channel and a program guide that listed close to 50 subjects from which to choose. They simply called up the particular item they wanted to see by dialing a special number and code on their tele-

Shopping at home with The CABLESHOP

phones. A computer programmed their request so that it would come up on one of three CABLESHOP information channels within the following few minutes. The segments included what to look for in buying a car (sponsored by an automaker), new recipes and specials at the local supermarket (from a food manufacturer), and information on how to save and/or invest money. As people used their sets to request information, they were no longer passive viewers, but, rather, active video consumers.

For advertisers, The CABLESHOP offers the chance to test and measure results for selling a product or service directly, generate qualified sales leads, build in-store traffic, provide couponing and sampling, and provide support for print and direct mail advertising.

A service dedicated solely to viewer and advertiser interaction and involvement is the UTV Cable Network. From 3:00 PM to 3:00 AM daily, UTV plans to feature shop-at-home, phone-in, opinion polls and games. For the home shopper, products will be displayed or modeled, with price details and an 800 toll-free response number flashed on the screen. In a home Bingo show, viewers play and phone in their winning cards for prizes when they win. In an instant millionaire game, the weekly game show winners will meet at the end of the year and one will become an instant millionaire. From "help-wanted" to "how-to," the emphasis is totally on response—playing along, phoning, mailing in, learning or buying.

Besides the 800-number facility, UTV has a computer-phone capable of handling over 1,000 calls a minute. Advertisers can thus have instant response to their messages. In addition, a sophisticated computer device allows national advertisers to tag names of their local

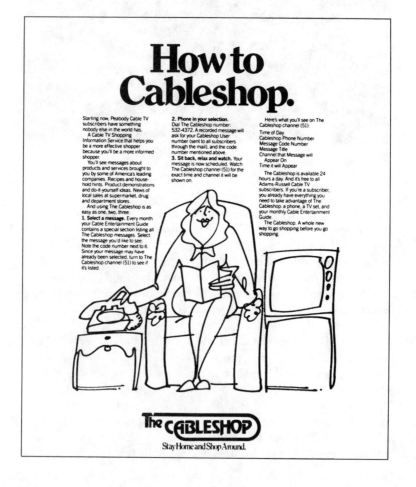

retailers or distributors in 5,000 different parts of the country. UTV will provide the local cable company with the terminal decoder needed to plug in the local information. In a nutshell, UTV, like The CABLESHOP, provides interactivity *without* interactive cable equipment.

An Issue of Support: Who Pays?

In addition to offering diversity of choice and time and creating an active viewer environment, video of the 1980s differs from television— and, in fact, from all major media—in its source of financial support.

The radio, television, and outdoor communications media all depend solely on advertising revenues for economic survival. Magazines and newspapers require advertising to exist, although a significant contribution to their financial well-being comes from newsstand and subscription revenue.

The new electronic media are quite different. Their economic health is dependent on consumer financial support as reflected in subscription fees and the purchase of a variety of video hardware and software. Advertising has a definite role to play in the economics of the new media, but in many instances it is a supplemental role, adding to their profitability rather than creating it. Individual cable satellite networks and certain interactive services will require advertising support to exist. But cable, video cassette recorders, video disc players and other video media overall will survive and prosper with or without it. If advertisers recognize this, they will perhaps be better able to develop advertising strategy and creative approaches for video of the 1980s.

All of the major broadcast and print media will be impacted by the new electronic media. We will focus on this in the next chapter.

CHAPTER 4

The Impact of the "New" Media on the "Old" Media

Since the word *cable* is generally followed by the word *television*, many advertisers tend to regard it as merely an extension of TV—like network television, spot television, and therefore, "cable television." In its earliest days, when it served only to improve reception and bring television to areas that otherwise could not receive the signals, cable was in fact merely an extension of television. Today, however, cable has a range of programming and services all its own.

Cable is not simply television with smaller numbers than network and spot TV. It is a new and distinct communications medium that offers high viewer involvement and selectivity and is, in many ways, closer even to many other media than it is to television. Cable is cable—or cable video, if you will. And just as it is important to consider its impact upon television, it is equally important to consider its potential effect on radio, newspapers, magazines, and out-of-home media.

Television

The moment an individual has the capability of receiving cable television, pay television (either cable or over-the-air), or has some form

of video recorder or playback device, his conventional television viewing patterns become altered.

Audience Levels

Nielsen studies have found that:

- Cable, particularly pay cable, leads to increased television
  usage.
- During prime-time and daytime, network viewing levels are
  cut into among pay cable homes.
- During the early fringe hours, network viewing is significantly
  lower in pay households than in noncable homes, *but*
- Pay cable seems to encourage more viewing (including viewing of network stations) during the late night hours.

While little data on the impact of over-the-air subscription and
MDS television are yet available, it does appear that pay movies and
special events transmitted in this manner are taking their audience toll
in many major markets where they are available. And, in homes with
any of the "time shift" media, viewing patterns are flip-flopped in a
variety of ways as consumers buy or rent prerecorded programming or
tape broadcast programming to watch at a later hour or date.

The new media will affect the advertising and marketing capabilities of television as more and more households become so equipped. By the end of this decade:

- Cable television will be in nearly 60 percent of the nation's
  homes.
- In more than 40 percent of the homes in the country, pay TV
  will be regularly viewed via cable, over-the-air transmission,
  or direct broadcast satellites.
- Video cassettes and discs will be in 20 percent of the households, letting people watch what and when they want.
- In more than one out of every 10 households, viewers will
  interact with their sets via Videotex services.
- And in a quarter of all U.S. homes, video games will be a part
  of everyday entertainment while computers will play a role in
  20 percent of all homes.

Since viewers will be doing more and different things with their
television sets, the levels of homes using television will increase. In

prime-time hours, for example, homes using television will increase from 60 percent in 1981 to 64 percent by the end of the 1980s. Network shares of audience will, however, decline from 82 percent to 61 percent. Thus, network television will remain the most mass of mass media, but the average program rating will have dropped from a 16 to a 13 by the end of the decade.

Of equal importance is the fact that the declines will not be uniform across all groups, programs, and times of the year. Access to the new media can be expected to climb much more rapidly among the better educated, more affluent, younger, and more innovative families. These are the most desirable advertising prospects. Live television—sports, news, and the most popular shows—will remain strong. "Marginal" shows—what Paul L. Klein, a former executive vice president for programs of NBC-TV referred to as the Least Objectionable Programs—will suffer most. They will not be all that a person has available to view. There will be Most Attractive Alternatives. And, similarly, during the increasingly long "rerun season," there will be new electronic video choices and even greater network audience declines.

Commercial Flipping and Zapping

While most television research has focused on how cable will affect broadcast audience levels, perhaps a more important area of concern is how it will affect advertising attentiveness. In buying radio to reach teenagers, it has been noted that when the news comes on a contemporary music station, kids will quickly flip to the nearest alternative where the rock beat is still blaring forth.

With cable, this same phenomenon will occur as the quantity and variety of programming increases *and* as there are more hand-held converter boxes in use. Television viewers, for example, know that network movie breaks are about two minutes long. They also know that between the end of one show and the beginning of the next lies about five minutes of advertising and other nonprogram material. The result: when a broadcast commercial break occurs, viewers will be able to quickly shift to CNN for an update on the news, or to ESPN for the latest in sports, or to The Weather Channel, or to a home shopping channel—or to any of a score of other channels with ease and with absolutely no compunction about skipping an advertiser's commercials. Figure 4–1 shows how viewing patterns may change.

There already are reports of cable subscribers switching from show to show as fast as they can push a button or flip a switch. And, in

Canada, where cable stands at about 70 percent penetration, children play a game called "zapping." The object is to see how long they can watch television without having to sit through a commercial!

In homes with video cassette recorders, flipping and zapping take on an added twist. For example, viewers can tape the early evening or late night news, and then play it back and fast-forward past stories they have little interest in—or commercials. The viewer chooses what he or she wants to see just as a magazine reader flips past articles and advertising he or she chooses to ignore.

If marketers fail to consider these ways in which the new electronic media affect advertising attentiveness, they naively assume that as cable and time shifting increase in impact, television's commercial impact will remain unchanged. (See Figure 4–2.) In fact, the pattern looks more like Figure 4–3. Or, perhaps, if advertisers and broad-

---

FIGURE 4–1. Commercial Flipping and Zapping

Movie	Commercials	Movie

CNN
ESPN
The Weather Channel
Home Shopping
∞

FIGURE 4–2. Unchanged Viewing Patterns

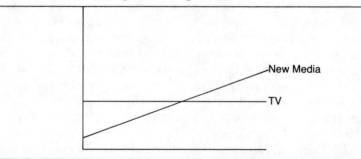

casters respond to flipping and zapping by making changes in commercial formats and by making the commercials themselves more interesting to hold viewer tuning and attention, the pattern may even look like Figure 4–4.

Radio

Like radio, cable is a highly localized medium. In fact, it is even more sharply localized. A cable system's boundaries can be precisely defined, and, unlike broadcasting, they are usually smaller and more manageable in size. Because of this, cable offers advertisers a new electronic media form in communities where they have relied mainly on radio (because television was either unavailable or unaffordable). Cablecasters in some markets are even considering joining forces with

FIGURE 4–3. New Viewing Patterns

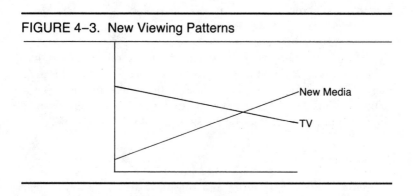

FIGURE 4–4. New Viewing Patterns

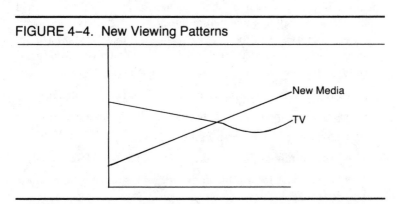

local radio stations in videocasting popular talk radio shows or disc jockeys. Sight would be added to sound in a "reverse simulcast."

Cable can provide many advertisers with the same low-cost, high-frequency, target-audience delivery they may now find in radio but not in television. For example, MTV: Music Television offers 24 hours a day of *video* music in *stereo*. It's a blend of TV and FM stereo radio that may feature artists singing their songs—or acting them out. Sometimes they are animated, sometimes they use experimental video art.

Advertisers who now use contemporary radio to communicate with youth or young adult target markets can deliver the same audience via MTV: Music Television on both a national network and (through local availabilities) on a local community basis. Chevrolet, for example, ran the first long-form, 90-second stereo commercial on MTV: Music Television for Camaro. It was an impressionistic montage set to a soft rock tune. A series of fast clips included body builders, a street sign reading Z28 (the car's model number), and a shot of the Camaro cruising past a desert sunset.

In still another area, cable satellite transmission is bringing new radio stations to markets that formerly did not receive them. Chicago's classical FM station—WFMT—is transmitted across the country as a "Super (radio) station," the radio equivalent of Ted Turner's Atlanta Superstation (WTBS), WGN (Chicago), and WOR (New York).

Newspapers

In many small communities and suburban areas, the only local medium is the suburban or shopping newspaper. Cable video will change this, providing a new selling medium for retailers, local dealers, and franchises. For the people in the community who have depended exclusively on their local paper, cable will deliver a wide array of community news:

- programs for and about local schools and senior citizens.
- programming from local business and civic organizations.
- coverage of village board meetings.
- programming from the community library and local colleges.
- coverage of local athletic events, celebrations, parades, and festivals.

All of these offer advertising potential, and recognizing that cable could cut into their local revenues, astute newspapers are quickly recognizing the value of the old adage "If you can't lick 'em, join 'em!" One such company is Leader Tele-Cable, which publishes the *Eau Claire (Wisconsin) Leader Telegram*. Not only does the company publish an electronic version of its 34,000 circulation daily, but it also has developed a classified ad channel that offers a full-color picture of an item for sale along with the price and seller's phone number. It complements the newspaper, but in an entirely different advertising form. We can also expect to see newspapers creating and cablecasting the news directly from their editorial offices, perhaps even joining together to provide a national satellite service.

One of the more imaginative alternatives to traditional newspaper advertising was developed by Televised Real Estate, Inc. Since 1979, it has occupied one channel on a local cable system in Spokane, Washington, and in September 1981, it began cablecasting on two systems in southern Orange County, California. For 17 hours a day, this real estate channel focuses on a variety of industry-related topics, but its main thrust is 60- and 90-second "commercials" for homes, clustered by communities. In each ad, the property is quickly toured and described, with special features highlighted. The end of each ad includes price and financing. Commercials for new home developments run to "infomercial" length of three to eight minutes. The real estate channel competes with the traditional newspaper real estate listings and can provide the incentive for a followup visit to the home.

Magazines

Magazines are embracing the new media avidly. Home Box Office carries *Consumer Reports*, *Ms.*, *Money*, and *Sports Illustrated* programming. *Better Homes and Gardens*, *Family Circle*, and *Scholastic* have developed cable shows, and *Playboy* plans to program an entire adult channel.

As the number of cable channels expands in the years ahead, viewers will find themselves browsing through video selections just as they now browse through the almost infinite variety of magazines at the newsstand. Already today, cable networks are devoted to:

- Arts and Culture
- Blacks

- Music
- News

- Children
- Consumers and Shopping
- Games
- Health

- Sex
- Spanish
- Sports
- Women

Ahead will be exclusive advertiser-supported channels focusing on magazine categories dealing with:

- Automobiles
- Beauty and Fashion
- Business and Finance
- Cooking and Foods
- Farming
- Gossip

- House and Home
- Literature
- Nature
- Senior Citizens
- Travel
- . . . many more

Many publications are working to develop videopublishing— programming for cassette and disc. These programs would encompass not only the consumer field but also could be developed to cover many business, industrial, and trade publications. Financial support would come from both subscriber and advertiser revenue.

An effective twist on videopublishing was the "Reader's Digest Do-It-Yourself Show" on the USA Cable Network. It was a 13-week series of entertaining and informative visual companion pieces to the "Reader's Digest Complete Do-It-Yourself Manual" and "Fix-It-Yourself Manual." a husband-and-wife team demonstrated how to build and how to fix things. The entire sequence of steps in each project was illustrated in color with computer-animated graphics.

It is not unusual to anticipate that the new electronic media will impact upon the magazine industry and that print will seek ways to interact with the new video forms. After all, magazines, cable, VCRs, and video discs have in common the attraction of an affluent and educated audience. The implication of this for advertisers is signifi-cant. Many marketers with highly "upscale" products have found that the audience appeal of broadcast television is too broad and have confined their advertising efforts exclusively to upscale magazines. Special-interest cable programs can extend these marketers' media options considerably and allow them to reach upscale audiences with video just as they do now with special-interest magazines.

One example is the increasing interest the public has shown in business and financial news, which has resulted in an expansion of business coverage in newspapers as well as the growth of national, local, and regional business publications. These, in turn, have been joined by a growing number of television and cable programs aimed at the same target:

- Wall Street Week (PBS)
- Business Journal (Syndication)
- This Week on Wall Street (CBN Cable)
- Financial Inquiry (CBN Cable)
- The Wall Street Journal Evening News (USA Network)
- Financial News Network (on television and cable)
- The Money Line (Cable News Network)
- The West Coast Report (Cable News Network)
- Money Week (Cable News Network)
- Inside Business (Cable News Network)
- The Business Channel (Manhattan Cable)
- Business Today (Satellite Program Network)

Advertisers seeking to reach a better-educated, upper-income audience through financial publications can add the video versions of these publications as a very logical market expansion opportunity.

Advertisers can also use cable to deliver a "unique editorial ruboff," just as they often do today in magazines. It is quite common for an advertiser to seek publications that have editorial subject matter that can reinforce its product message. Thus, a furniture manufacturer looks for magazines with home decorating and remodeling features, while a clothing manufacturer prefers to locate ads near fashion editorials. Such opportunities are available in cable, often in some very unusual ways.

A weekly series on the CBN Satellite Network is "Fresh Ideas with Claire Thornton." Each week's show includes a report on what produce is in season and is plentiful and a feature focusing on a particular aspect of the produce industry, such as a grower, a commission, or an association. Finally, Claire provides tips on how to select, store, and prepare fresh produce.

Several years ago, a leading salad dressing manufacturer examined magazines not only on the basis of the amount of food editorial they carried but also on the basis of the number of editorial items dealing with fresh vegetables and salads. "Fresh Ideas with Claire Thornton" provides the same kind of editorial environment normally confined to print. It is not unusual to find the show's advertisers including companies such as Sunkist, Dole Pineapples & Bananas, Washington State Apples, and the Imported Winter Grape Association.

Out-of-Home

No two media are as similar in geographic concentration of audience as are out-of-home and the new video. Out-of-home can profit from this. Outdoor locations within the community can specifically highlight the value of signing up for cable. By picking sites near theaters, pay services, VCR, and video disc manufacturers can reach people on their way to the movies and highlight the economy, convenience, and comfort of watching films in one's own home. Transit cards can include tear-off schedules of pay service offerings which include mail-in cards for further information.

The impact of the array of new media devices can actually create important new areas of revenue growth for out-of-home.

Media and the Future

What all of this means, in brief, is that the world of media will never be quite the same. It will provide new opportunities, new challenges, and new problems. Planning, in particular, will be radically different from planning in the past. That is the subject we will now turn to.

CHAPTER 5

The New Media Planning Process

The planning and execution of cable advertising strategy can be described as a classic case of trying to put the right round pegs in the right round holes in an environment where there is an equal number of square holes confronting the advertiser. The new era of video diversity presents as many problems as it offers opportunities:

- How does the advertiser know *where* to reach the best prospects?
- How does the advertiser know *how many* have been reached?
- How does this advertiser know what is the *cost* of reaching prospects, if indeed they can be reached?
- How does an advertiser know *what kind* of advertising can communicate most *effectively* with them?

In short, how does an advertiser piece together a cable plan when everyone is doing his or her own thing with his or her own television screen?

Having learned everything he can about all of the new cable video opportunities that are opening up, what does the advertiser do *now*?

The secret to success is: *Never evaluate a cable opportunity in isolation!*

Cable video is a medium that should be evaluated in the context of *all* of the media options available to an advertiser. (See the configuration in Figure 5–1.) The decision to use cable must be based on the fact that it offers a new or better way to accomplish an advertiser's marketing and/or communications goals.

What Can Cable Do for You?

Before deciding to use cable, an advertiser must carefully evaluate his existing media plans to determine how well they are accomplishing everything he would like them to do. An effective way to do this is to take the Advertising Media Quiz:

- I am basically satisfied with my advertising and with my media plan.
- It appears to showcase my product (or service) well.
- Research shows that it is conveying to my customers pretty much what I want to tell them.
- It seems to be efficient.
- I believe that it is doing an effective job.

I only wish that (fill in the blanks)

_____

_____

_____

_____

_____

_____

The answers will take a variety of forms, including:

- I only wish the media conveyed my messages in a more compatible environment and acted as more than just a commercial "carrier."
- I wish I could afford to sponsor something where I could get some real identity.

FIGURE 5–1. Cable Video—One of Many Advertising Options

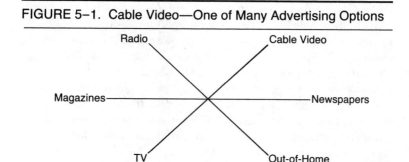

- I wish I had more time to tell my story. Thirty seconds is not enough, but that's all my television budget can afford.

- I wish I could have the advantages of television, but with the ability to say everything I now say in my print ads.

- I wish I could advertise somewhere that appealed more to the very special people that buy my product.

- I wish I could afford the high frequency of exposure I need but can now get only in radio.

- I wish I could zero-in more closely on some of the small towns where my dealers are located.

- I wish I could run a dozen or more different commercials, but I can't afford to spend all the money it takes to produce them for broadcast television.

- I wish I could experiment a bit and explore some new commercial ideas—maybe even try some direct response options.

Examining these responses can help an advertiser decide whether or not cable can provide a positive addition to the media mix.

Looked at in the context of all media, cable offers a variety of highly *specialized programming* that allows advertisers to zero-in on highly *targeted audiences* they might otherwise find hard to reach. Cable advertising is available at relatively *low costs per advertising unit*. This allows marketers who cannot afford broadcast television to take advantage of its visual and audio communications potential. Cable has the ability to provide *high frequency of exposure* to an advertiser's prospects throughout the day and offers *flexible message lengths* that meet advertiser communications needs rather than broadcast time limitations. Cable offers *program sponsorship opportunities* with advertiser identity. The program becomes more than just a commercial carrier. Cable provides *product or service exclusivity* in programs at affordable out-of-pocket expenditures. Cable audiences, in general, are *better educated, above average in income, employed in higher-level jobs, and younger.* Cable offers *opportunities to test creative ideas* at very low media costs. Cable provides many opportunities to tie-in advertising messages with *direct response offers.* Cable is *highly localized* and can provide advertising support for franchisees, dealer organizations, and wholesale and retail sales forces.

Looked at another way, cable offers: the visualization of television, the targeted formats and specialization of magazines, the low unit costs and frequency of radio, and the local interest and appeals of newspapers. (See Figure 5–2.)

Understanding Cable's Deficiencies

Cable is such a new medium that considerable judgment is required in evaluating how to use it most effectively. While cable has many assets, many of these are, at present, relatively unproven. Thus, in the early 1980s, cable as an advertising medium is perhaps at the same stage as broadcast television was in the mid-1950s.

Lack of Audience Information

A giant missing link in the cable advertising evaluation equation is a substantial body of audience research. Most people recognize that

FIGURE 5–2. Opportunities Offered by Cable

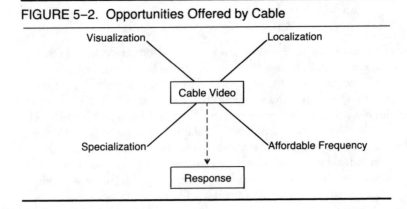

cable audiences are substantially smaller than broadcast audiences. However, the information that is available today comes in a variety of forms and a number of different bits and pieces. Furthermore, there is considerable lack of agreement within the industry as to just what to measure and how to measure it (more on that in Chapter 8). For some satellite network services, such as WTBS-Superstation and the Cable News Network, audience data are available on a consistent and continuous monthly basis from the A.C. Nielsen Company. For other cable networks and most local systems, such information is available only as occasionally ordered special analyses.

Spotty Coverage

In many major markets, cable's coverage is still very spotty. For example, in February 1982, 29 percent of the homes in the country were hooked up to cable. However, in 11 of the nation's top 20 television markets, less than 20 percent of the homes had cable. (See Table 5–1.)

Frustrations in Evaluating, Buying, and Creating Advertising

For many advertisers and advertising agencies, the multitude of cable offerings, uncertainties as to how to evaluate them, and frustration as to how to prepare special advertising for them have resulted in considerable confusion and anxiety. Agencies on a traditional 15 percent commission system are also finding themselves in a severe cost squeeze.

TABLE 5–1. Cable Penetration in the Top 20
U.S. TV Markets (Percent)

New York	24.4	Dallas–Ft. Worth	14.9
Los Angeles,		Houston	18.8
Palm Springs	19.8	Pittsburgh	45.4
Chicago	5.9	Miami–Fort Lauderdale	18.1
Philadelphia	29.4	Seattle–Tacoma	35.7
San Francisco–Oakland	40.4	Minneapolis–St. Paul	7.0
Boston, Manchester,		Atlanta	23.1
Worcester	18.1	St. Louis	8.3
Detroit	8.2	Tampa–St. Petersburg,	
Washington, D.C.,		Sarasota	22.1
Hagerstown	12.3	Denver	12.3
Cleveland, Akron	22.2	Baltimore	8.0

Source: A. C. Nielsen, February 1982.

The small dollar amounts involved in most of today's cable buys make it impossible for agencies to profit from them. For example, one network commercial on the 1982 Super Bowl cost $345,000. For this much money, a *very large* cable network *schedule* spanning many months could have been purchased. In fact, during the first six months of 1982, a major advertising agency purchased *300* commercials for *five* different clients on *six* satellite networks for $350,000. The cable buys were obviously several times more complex, costly, and time consuming to administer than the Superbowl buy. It is no wonder that agencies will find it necessary to ask advertisers for fees so they can afford to spend the necessary time to negotiate cable buys and produce special commercials for them.

Why Get Involved Now?

With cable's currently small potential audience size and the problems involved in evaluating, buying, and creating advertising for it, many advertisers ask, "Why not wait?"

The answer involves the "risk/reward relationship." Every informed source indicates that by the end of this decade, traditional television will have been transformed into a multitude of video offerings available to a majority of the population. The investment of time and money to explore, experience, and experiment today will better prepare the advertiser to effectively use these new media offerings tomorrow.

Those early pioneers who were attracted to broadcast television during the late 1940s learned what worked and what didn't work at a time when making a mistake carried a very small price tag. Thus, Kraft was the first advertiser to sponsor a weekly television drama, starting on May 7, 1947, in New York when there were only 32,000 TV homes in the entire city. Of course, what was learned then about producing a weekly program far outweighed the costs. Later, in 1956, they began producing all of their commercials in color, even though there were fewer than 200,000 color sets in the country. Again, Kraft knew that food should be shown in color and that color would eventually reach the majority of homes nationwide. The cost of learning how to use color properly in 1956 was far smaller than it would have been if they waited until years later, when the networks all began to telecast in color. The same holds true with cable. The rewards of learning how to best use cable today far outweigh the risks of procrastinating.

But the potential of cable is not just in the future. Many opportunities exist to use cable efficiently and effectively today. The key is to select alternatives and then implement those that will best achieve a set of advertising objectives, whether they be for the Ford Motor Company or for a Ford dealer in Little Rock, Arkansas, for American Express or for a small bank in the suburbs of St. Louis.

Taking Advantage of Publicity and Promotion

Through the creative use of well-planned and well-timed publicity efforts, the impact of today's buys can be increased and considerably extended. Publicity can promote the cable effort among cable subscribers to increase viewership and announce the advertiser's use of cable to the trade, business leaders, government and community leaders, educators, the press, and other important influences affecting the advertiser's business.

Since many people will not have access to cable, the judicious use of print advertising can announce a cable campaign to many who would otherwise be unaware of it. For example, an ad in the *Wall Street Journal* announced that Kraft had signed up as the first advertiser on CBS Cable. It reached an important group of thought-leaders who would not otherwise have been aware that the Kraft Music Hall was returning—this time to cable.

It is impossible to spell out in detail all of the guidelines for effective publicity; all of the traditional approaches certainly apply. Not only can a cable program be promoted by an effective publicity effort, but publicity efforts themselves can often be turned into well-executed cable advertising programs. For example, many high schools across the country offer evening how-to-fix-it courses in conjunction with a local hardware store or lumber dealer. A natural extension of this would be the production of a series of how-to programs by the community's local cable system. These would be funded by the participating hardware store or lumber dealer and would feature their products in use. Thus, what was simply a public relations effort could become an effective cable advertising program.

A very successful prototype for this kind of program is already in existence. The Public Broadcasting System's "This Old House," which originates from WGBH in Boston, is devoted to the how-to approach. Specifically, the program shows how to handle nearly every conceivable remodeling problem so that an old house can be completely refurbished and made ready for sale or habitation. At this point

This ad announced the return of Kraft Music Hall, this time on CBS Cable.

the program's host is working on his third "old house" and has produced a series of books based on the program.

Establishing Cable Advertising Objectives

Effective cable advertising executions begin first by answering the same series of questions you answer in evaluating magazines, news-papers, out-of-home, radio, and television:

- *Who* are the people with whom you want to communicate?
- *Where* do you want to reach them?
- *When* do you want to reach them?
- *What* information do you want to communicate to them?
- *How* many do you want to communicate with, *how* often, and *how* much do you want to spend to do it?

The answers to these questions establish your cable advertising objec-tives.

Next you need to establish a cable strategy by examining the degree to which cable can meet your objectives. The "Checklist of 'New Media' Media Values" (Table 5-2) provides a framework for this examination and strategy. It lists 18 media evaluation values to be considered and shows where these values are found in the "Non-New Media"—magazines, newspapers, out-of-home, radio, and television. Finally, it shows where these same values exist in the "New Media." Included are:

- Cable
  - Satellite Cable Networks
  - Regionally Interconnected Cable Systems
  - Individual Cable Systems
- Time-Shift Media
  - Video cassette Recorders
  - Video disc Players
- Information Systems (two-way Cable, Videotex)

FIGURE 5–3. Developing Strategy for Selecting and Using the New Media

Identifiable Advertising Objectives	+	Identifiable Cable Opportunities	=	Intelligent New Media Usage

Finally there is the evaluation of the individual cable opportunities (the programs, the networks, and the cable systems) and then the actual buy and all of the subsequent follow-up. Figure 5-3 diagrams this process.

TABLE 5–2. A Checklist of "New Media" Media Values

Media Evaluation Values	Source of Values in "Non-New" Media					Opportunities in "New Media"				
						Cable				Info.
	M	N	O	R	T	Sat.	Int.	Loc.	VCR Disc	Sys-tem
Who										
Audience Selectivity	X			X	X	X	X	X	X	X
Upscale Audience Profiles	X					X	X	X	X	X
What										
Visibility, Sound and Action					X	X	X	X	X	
Product Enhancing Environment	X					X	X	X	X	X
Product Demonstration Potential					X	X	X	X	X	
Flexible Ad Message Length	X	X				X	X	X	X	
Newsworthy Setting		X						X		X
Direct Response Generator	X	X				X	X	X	X	X
Quality Color Reproduction	X		X			X	X	X	X	X
When										
Fast Dissemination of Information		X		X	X	X	X	X		
Long Life	X		X						X	
Short Notice for Late Buys		X		X	X	X	X			
Where										
Geographic Area Targeting	X	X	X	X	X		X	X		X
Highly Localized		X	X	X	X			X		X
Retail or Dealer Tie-In Potential		X				X	X	X	X	X
How (Many, Often, Much)										
High Reach Potential		X		X						
High Frequency of Message Delivery				X	X	X	X	X		
Low Unit Costs					X	X	X	X		

M – Magazine   R – Radio   Sat. – Satellite
N – Newspapers   T – Television   Int. – Interconnects
O – Outdoor   Loc. – Local

Implementing a Cable Advertising Program

In March 1981, the American Association of Advertising Agencies surveyed all offices of its member agencies to determine their procedures for buying cable television. At that time, one-third of those who responded indicated that their offices had been involved in a cable buy. Among those buying agencies, different individuals were responsible for making the buy. (See Table 5-3.)

While the specific execution of a cable buy is a media function, the most effective cable advertising comes about through a very close working relationship among everyone involved in the buying and selling of cable. This includes all those involved in evaluating it from a media standpoint, from a research standpoint, and from a creative standpoint.

TABLE 5–3. Who Buys Cable?

	Percent of All Buys	Breakdown by "Type" of Buy		
		Local	Regional	Satellite
Spot Buyer	30	65	20	15
Network Buyer	19	—	6	94
Media Planner	37	43	6	51
Special Cable Buyer	6	20	10	70
Other	8	N/A	N/A	N/A
	100			

Source: AAAA, "Results of Survey on Cable Buying-Paying Procedures," August 1981.

FIGURE 5–4. Making a Good Cable Buy

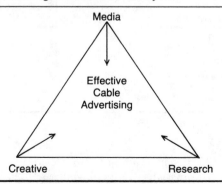

Getting the Job Done

The setting and implementation of a sound cable advertising program follows an organized step-by-step procedure. Figure 5-5 and following figures show this process in a flow-chart format.

Step One

Examine your product or service in terms of its overall media and communications requirements. This will provide the basis for determining how well each cable opportunity can accomplish your objectives.

FIGURE 5–5. Setting and Implementing Cable Advertising Strategy: Step One

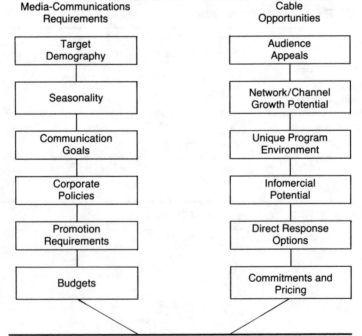

Media-Communications Requirements	Cable Opportunities
Target Demography	Audience Appeals
Seasonality	Network/Channel Growth Potential
Communication Goals	Unique Program Environment
Corporate Policies	Infomercial Potential
Promotion Requirements	Direct Response Options
Budgets	Commitments and Pricing

Step Two

In the negotiation stage, you will evaluate the specific cable proposal(s), negotiate the advertising schedule, and attempt to build in some form of long-term price protection to protect against sudden escalating costs. (See Figure 5-6.)

FIGURE 5–6. Setting and Implementing Cable Advertising Strategy: Step Two

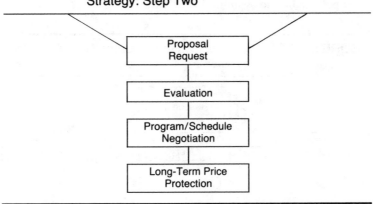

Step Three

The order is placed. Now it is time to prepare special commercials, plan publicity and promotion, and design any special research to measure the effectiveness of your cable schedule (Figure 5-7).

FIGURE 5–7. Setting and Implementing Cable Advertising Strategy: Step Three

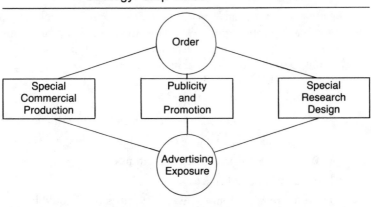

Step Four

Now, evaluate the response to your cable advertising either through audience research, direct response offers, or any special research you have designed. At the same time, you should continue to explore all other potential cable opportunities (Figure 5-8).

FIGURE 5–8. Setting and Implementing Cable Advertising Strategy: Step Four

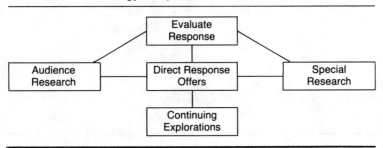

This planning strategy applies to local as well as to network cable. Advertisers can use it to direct their buying activity, and cable systems and networks with an understanding of it can do a better job of selling the medium. Figure 5-9 fully illustrates this process.

Achieving Cable's Maximum Benefits: Case Histories

Throughout the entire cable planning process, it is essential that an advertiser "keep on the right track." This involves a complete examination of what you want cable to accomplish, a thorough investigation of all cable opportunities available to do this, and a constant re-examination of what is being done.

In a nutshell, one must ask and re-ask:

1. What is our advertising media strategy accomplishing?
2. What would we like it to accomplish that it is not now doing?
3. What cable opportunities can help us do this?
4. Is our cable effort succeeding in doing this?
5. Are there additional cable opportunities that should be considered?

FIGURE 5–9. Setting and Implementing Cable Advertising
Strategy

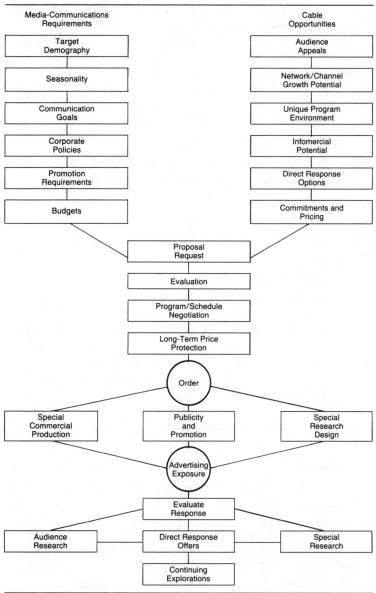

The remainder of this chapter is devoted to brief case histories of companies that have followed the planning process described above. Although they are very different companies marketing very different products, they have at least two things in common. One, they identified their objective, thoroughly investigated ways of attaining that objective, and reexamined their strategy. Two, all were successful to the extent that they gained important new knowledge about a new medium at a price that will never be lower.

Advertisers have used cable to achieve any one or a combination of several communications benefits. Following are 15 case histories of how cable was used by different companies to deliver or secure:

- A specialized, targeted program environment.
- A flexible commercial in terms of length or form.
- Program sponsorships and product exclusivity.
- Direct response.
- A low-cost testing potential.

Table 5-4 shows how the advertisers in these cases accomplished these objectives.

Anheuser-Busch

Goal: To test the ability of television to sell its Clydesdale Collection, a mail order catalog line of etched crystal mugs and beer glasses, home entertaining gifts, and personal accessories. Previously, the Clydesdale Collection's advertising had been confined solely to magazines.

Cable Idea: A special 90-second cable advertising message was created for exposure on ESPN sports programming. Its aim was to increase product awareness, catalog requests, and sales. Cable permitted television to be tested at low media and commercial production costs.

Hallmark

Goal: To extend the reach and promotional impact of the "Kaleidoscope" show, which is staged at its headquarters in Kansas City and taken on tour to other cities. The show centered on crafts— how to make things and how to put things together.

Cable Idea: Hallmark produced a series of five-minute programs directly related to "Kaleidoscope." They aired them on the USA Network's "Calliope" to reach children across the country.

TABLE 5–4. Achieving Cable's Maximum Benefits

	Anheuser-Busch Clydesdale Collection	Hallmark Kaleidoscope	Kawasaki and Music Television	Kemper Insurance Co. Golf	Ken-L Ration Pet Show	Ken-L Ration and Westminister	Kraft, Cookbooks and Recipes	McDonnell-Douglas DC-10	Old Spice Sports Quiz	Quaker Oats and CBS Cable	Raisinets and the Movies	Scott Value Line	3-In-1, Plastic Wood and I.D.s	20th Century Fox Promos	Wilton Cake Decorating
Specialized-Targeted Programming	X	X	X	X	X		X	X	X	X	X	X		X	
Flexible Message Lengths and Form	X		X		X			X	X	X	X	X	X	X	
Program Sponsorships		X		X	X	X		X		X					
Product Exclusivity		X		X	X	X				X		X			
Direct Response	X			X	X	X	X	X			X			X	
Low Cost Testing Potential	X					X		X			X			X	

Kawasaki

Goal: To create a low-cost, unique commercial message targeted specifically at Kawasaki's young audience and closely associated with the style and feeling of the media vehicle for which it was designed.

Cable Idea: A two-minute "musical film," with an original music score written especially for MTV: Music Television, was created. Kawasaki motorcycles, of course, were featured. The production involved the re-editing of existing commercial elements—motorcycle film and stock footage—and the specially written rock music was based on the "Kawasaki Lets the Good Times Roll" theme.

Kemper Insurance Company

Goal: To provide additional advertising and promotional support for the annual Kemper Open Golf Tournament on the CBS Television Network.

Cable Idea: Each year, Kemper produces a special golf film based on the current year's tournament. The film is used by Kemper agents and made available to local clubs and organizations as well as to television stations.

Prior to the 1982 tournament, the 26-minute 1981 Kemper Open film was run three times on ESPN. It created a greater in-depth awareness of the event than could have been conveyed in shorter broadcast length tune-in announcements.

Ken-L Ration

Goal: To enhance Ken-L Ration's commercial message environment, strengthen its image among consumers, foster good will among veterinarians, and broaden distribution of its consumer promotional material.

Cable Idea: Ken-L Ration appeared daily in the Cable News Network's "All About Pets," a 2-3 minute program segment covering topics on the care and feeding of pets.

A 15-second slide tag was added (at very little production cost) to Ken-L Ration's 30-second broadcast commercials. It offered Ken-L Ration's booklet on "How to Care for, Train, and Feed Your Dog" and reminded viewers about the importance of visiting their veterinarians regularly.

Ken-L Ration

Goal: To showcase Ken-L Ration in a unique environment related to dogs and targeted specifically toward dog owners.

Cable Idea: Sponsorship of the USA Network's six-hour coverage of the Westminister Kennel Club's Dog Show. The most prestigious and famous dog show in the country, it offered Ken-L Ration:

- A large concentration of dog owning viewers.
- A program environment geared 100 percent to the product.
- Exclusivity as the only dog food advertised in the event.
- Favorable publicity in the form of press releases in trade publications.

Kraft

Goal: To gain additional exposure for commercials whose previous airings had been limited to network television specials *and* to provide an opportunity for viewers to respond to write-in offers for Kraft recipes.

Cable Idea: The audio for "Baked Goods from Around the World," a 90-second Christmas holiday special, was retracked to offer a free recipe booklet by mail.

Following this, elements of several commercials were re-assembled and re-tracked into a 90-second cable message offering Kraft's 75th Anniversary Cookbook for $4.95 to those viewers who wrote in for it. These commercials ran on the Cable News Network and WTBS.

McDonnell-Douglas

Goal: To boost the impact of its DC-10 advertising effort among key governmental and military decision makers as well as the public at large.

Cable Idea: On ESPN, McDonnell-Douglas sponsored the annual Army-Air Force football game. Its commercials were enhanced by the addition of a slide tag that offered a booklet with further information on the DC-10.

Old Spice

Goal: To create a compatible environment for this men's fragrance and determine (without existing rating data) which of several sports programs offered it the greatest audience attraction.

Cable Idea: Old Spice created a sports trivia contest that centered on events and stars in baseball, basketball, boxing, football, and hockey. The commercials ran in a variety of ESPN sports programming with each program's responses identified by a different code. Thus, Old Spice developed its own measure of audience appeal without waiting until rating data were available on ESPN.

The Quaker Oats Company

Goal: To demonstrate Quaker's commitment to cultural arts programming and reach an upscale target audience in new, interesting, and creative ways.

Cable Idea: Quaker purchased full sponsorship of CBS Cable's entire October 12, 1981, premiere night of programming. From a creative standpoint, Quaker blended its existing food and pet commercials with specially created variable length messages that:

- Explained Quaker's dedication to quality, all-family programming.
- Described the care and expertise that went into the design and building of Fisher-Price toys.
- Told the story of a boy and his little lost dog.
- Presented the evolution of the Quaker logo.

Raisinets

Goal: To reach an under-30-year-old audience, which represents the primary consuming group, and to increase its association with movie going, where a large volume of the candy's consumption takes place.

Cable Idea: A schedule of 10-second and 30-second announcements were run on MTV: Music Television. The 30-second commercial was a humorous parody of movie scenes and a play on words in movie titles. It ran immediately adjacent to movie company spots on MTV: Music Television and created a complete movie environment for the product. The 10-second message featured the name *Raisinets* in a visual of a theater marquee sign. This commercial highlighted the fact that Raisinets are available in your local theater, and it then led into a theatrical movie commercial.

Scott Paper Company

Goal: To develop an advertising feature that could promote Scott products in an environment that demonstrated the company's service orientation to the consumer.

Cable Idea: Scott created the "Scott Value Center" for the Modern Satellite Network's "Home Shopping Show." Running twice a day, five days a week, these one-minute tips provided helpful information, promoted Scott's wide variety of products, and offered viewers booklets and product samples.

3-In-1 Oil and Plastic Wood (American Home Products)

Goal: To secure national exposure for a group of 10-second commercials that could only be aired as newsbreaks in network television or in local-market spot positions.

Cable Idea: The commercials were placed in male-oriented sports programming on ESPN and the USA networks. These cable networks had the flexibility to accept shorter length spots as well as long-form messages.

Twentieth Century-Fox

Goal: To promote the release of a new film, "Fort Apache, the Bronx," with more than the usual 30-second network television commercials.

Cable Idea: The full two-minute theater trailer with a 5-second tag ("Starts February 6 at theaters everywhere") was run on CNN, ESPN, and WTBS.

Wilton Enterprises

Goal: To provide a greater ability to demonstrate and explain Wilton's cake-decorating products and courses than was possible with its 30-second broadcast commercials.

Cable Idea: Wilton developed and aired a 9-minute segment for the "Home Shopping Show." It presented cake-decorating techniques and concluded with a beginner's book offer and a toll-free 800 number for further information.

CHAPTER 6

Different Avenues for Different Advertisers

The Scope of the Business: National or Local

All advertisers are not created equal. As a result, cable opportunities should be evaluated initially in relationship to the scope of each advertiser's business and communications objectives. Is the advertiser's business national or local? Is the advertiser's product or service of mass use and appeal, or is it more narrow? The answers to these questions provide the key to the successful use of cable as an advertising medium.

National Satellite Networks

Three avenues are open to the prospective cable advertiser. The advertiser-supported national satellite networks provide a variety of options, some that have relative mass appeal and others that are narrower in viewer selectivity. While the term *national satellite network* implies uniform coverage throughout the country, such is actually not the case. Until the larger metropolitan markets are wired, there still will be substantial soft spots. For example, as of February 1982, 29 percent of the nation's homes were cable. But, in many of the country's top markets, it was well below that level. (See Table 6-1.)

The absence of significant levels of cable coverage in many large markets should not deter an advertiser from using the satellite-fed cable services *unless these markets represent primary areas of sales strength.* Despite limitations of coverage in some areas, the number and variety of advertiser-supported satellite program services are continually increasing. The only way to be certain about current advertising opportunities is to check updated sources such as the Cabletelevision Advertising Bureau, the TAG Cable Information Exchange, the cable trade press, and the satellite networks. Table 6-2 lists the major advertiser-supported satellite services as of July 1982.

Interconnects

Cable interconnects represent several cable systems in a given area that join together to exchange cable programming and to sell advertising with the convenience of one-order, one-bill placement. They can be connected either electronically via a microwave hookup or by "bicycling" videotapes between participants.

Interconnects allow a national or local marketer to use cable to both add and replace other advertising weight going into an area. While an individual cable system might cover only a portion of a market, an interconnect may represent the equivalent of an entire ADI, DMA, or Standard Metropolitan Area. Carl Weinstein, president of Eastman CableRep, which sells advertising time on a number of interconnected systems, refers to them as CMOs—or Cable Markets of Opportunity—to distinguish them from broadcast television markets.

TABLE 6–1. Cable Penetration in the Top 20
U.S. TV Markets (Percent)

New York	24.4	Dallas–Fort Worth	14.9
Los Angeles,		Houston	18.8
Palm Springs	19.8	Pittsburgh	45.4
Chicago	5.9	Miami–Fort Lauderdale	18.1
Philadelphia	29.4	Seattle–Tacoma	35.7
San Francisco–Oakland	40.4	Minneapolis–St. Paul	7.0
Boston, Manchester,		Atlanta	23.1
Worcester	18.1	St. Louis	8.3
Detroit	8.2	Tampa–St. Petersburg,	
Washington, D.C.,		Sarasota	22.1
Hagerstown	12.3	Denver	12.3
Cleveland, Akron	22.2	Baltimore	8.0

Source: A. C. Nielsen, February 1982.

Table 6-3 lists major interconnects either operating or being started up as of mid-1982.

The extent of an interconnect's coverage area can stretch for a considerable distance. For example, Gillcable serves 85,000 homes in its own franchise area in Santa Clara County and reaches an additional 350,000 in the San Francisco area from Napa in the North to

TABLE 6-2. Advertiser–Supported Satellite Services

Satellite Program Service	Programming	July 1982 Subscribers (Millions)
ARTS	Primetime visual and performing arts	7.5
Black Entertainment Television (BET)	Movies, sports, specials	3.0
Cable Health Network	24-hour health, diet, nutrition and medical	4.6
Cable News Network I (CNN I)	24-hour in-depth news and information	14.0
Cable News Network II (CNN II)	24-hour short form news segments	2.0
CBS Cable	Evening cultural and performing arts	4.3
CBN Satellite Network	24-hour all family programming	17.0
Daytime	Weekday women's service	5.5
ESPN	24-hour sports	17.0
Modern Satellite Network (MSN)	Daily informational/educational	5.0
MTV: Music Television	24-hour video music	5.2
Nashville Network	Daily country music and variety	*
Satellite NewsChannels I	24-hour short form news segments	2.6
Satellite NewsChannels II	24-hour long form news and analysis	*
Satellite Program Network (SPN)	24-hour features, information and entertainment	5.0
Spanish Entertainment Network	24-hour Spanish programming	2.7**
USA Cable Network	24-hour women's, children's, sports, and entertainment	12.0
UTV: Involvision	Daily viewer involvement and participation	*
Weather Channel	24-hour weather	3.0
WGN (Chicago)	24-hour movies, entertainment and sports	9.3
WOR (New York)	24-hour movies, entertainment and sports	5.5
WTBS Superstation (Atlanta)	24-hour movies, entertainment and sports	23.2
*Scheduled for Launch	**Spanish-speaking homes	

Salinas in the South. And the Philadelphia interconnect encompasses systems in Pennsylvania, New Jersey, and Delaware.

From the marketer's standpoint, an interconnect permits an advertising message to be targeted to the specific demographic or geographic audience he wants to reach. Commercials can automatically be slotted in the local advertising positions of cable satellite networks (Cable News Network, ESPN, CBS Cable, MTV: Music Television, etc.) and other programming carried by the individual cable systems in the area. The advertising can be transmitted to all of the interconnect's cable systems, or, in the case of the more highly developed interconnects, it can be directed only to those specific communities the advertiser wants to reach. In this case, a lawn care product might have its message directed only to systems in areas with a high incidence of single-family dwellings.

Looking ahead a few years, super-interconnects may be developed to serve larger regional areas, such as the entire West Coast. The technology is already available, but advertiser interest will determine how soon this is accomplished.

Strictly Local Cable

Unlike broadcast television, where a station's economic survival depends on the advertising dollars it generates, cable's primary source of income comes from subscriber revenue. Thus, the majority of the nation's 5,000 cable systems have yet to take an aggressive stance in selling advertising, and the best estimates are that only between 15 percent and 20 percent of them are now doing so. This number, however, is rapidly rising as system owners have begun to see the potential profits that advertising can generate.

To date, most advertising support of cable has been with the

TABLE 6–3. Cable Markets of Opportunity

Atlanta	Philadelphia
Boston	Pittsburgh
Buffalo–Rochester	San Diego
Cleveland	San Francisco Bay Area
Connecticut	Seattle–Tacoma–Spokane–Portland
Iowa	Syracuse–Binghamton–Utica–Rome
New York Metro	Wichita–Hutchinson
Palm Springs	

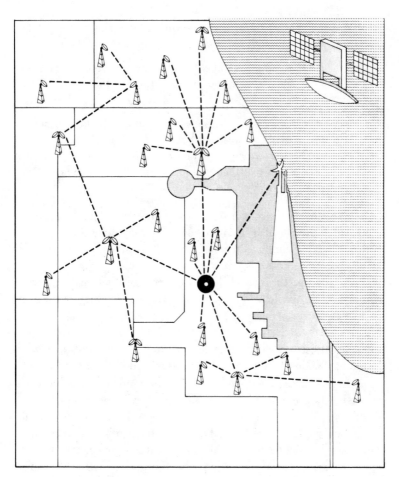

A proposed Chicago-area interconnect

national satellite networks. Its real, long-term potential, however, may be more at the local grassroots level. And, unlike broadcast television from Los Angeles, Omaha, or Atlanta, cable advertising will impact far differently on the cities and suburban communities it serves. It offers several distinct benefits to advertisers attempting to target local audiences:

- Since cable systems have carefully defined boundaries, the retail points affected by advertising can usually be identified.

- An advertiser knows the area where advertising is being delivered, and there is no "spill-out" or "spill-in."

- The cable family is directly tied-in to the cable operator and receives at least one monthly mailing—a bill and/or a program schedule.

- With cable's many channels, there should be no shortage of time, as with network and spot television. As a result, an advertiser can expect to hold the line on costs and can expect greater flexibility in scheduling and in commercial design and length.

In a large metropolitan area, advertisers already have a full media menu to satisfy most of their needs. Chicago, for example, has 7 commercial TV stations, 50 radio stations, a complete complement of general and ethnic newspapers, a city magazine, 4,300 billboards, posters, and car cards, and dozens of localized national magazine editions. Here, cable won't so much represent a totally new medium as it will a means of delivering a unique creative treatment that existing broadcast television does not provide for. This includes in-depth personal selling, advertising that is tailored to a specific program's environment, and affordable sponsorships as existed in the 1950s.

In the smaller towns and suburban communities, however, cable is a brand new communications form. In those areas, a franchisee, a dealer, or a retailer's media options have been largely limited to suburban newspapers, possibly a radio station, billboards, and direct mail. Television has not been practical either because its price tag or because coverage extended outside of an advertiser's area. Cable, however, offers an attractive alternative or addition to the local advertising media mix—low-cost, informational, service-oriented communication tailored to a specific group of prospects.

By 1990, well over half of the nation's homes will be hooked up to cable. For the advertiser looking for the heaviest pockets of cable penetration today, Table 6-4 shows those Nielsen Coverage Areas that have already joined the 1990 "50%-Plus Club." In addition to these DMA's, there are many smaller communities and suburban areas with high levels of cable penetration and interest in attracting advertising revenue to their systems.

The Nature of the Product or Service: Mass-Market or Non-Mass-Market

Most mass-marketed goods and services are relatively similar in use and in target audience. As a result, advertising has generally sought to differentiate them from a creative standpoint by developing distinct brand personalities and focusing attention on what Rosser Reeves referred to as the Unique Selling Proposition.

In terms of media selection, the differences in plans between competing products and services are often quite small. In radio and

TABLE 6–4. Nielsen Coverage Areas with over 50 Percent Cable Penetration

Santa Barbara–Santa Maria–		Biloxi–Gulfport	59.8%
San Luis Obispo	79.2%	Amarillo	58.8
San Angelo	76.5	Ft. Myers	58.5
Laredo	75.8	Gainesville	58.4
Marquette	75.4	Salisbury	58.3
Parkersburg	73.8	Glendive	57.2
Casper–Riverton	69.9	Bend, Or.	56.7
Johnstown–Altoona	68.8	Eugene	56.4
Clarksburg–Weston	68.8	Ft. Smith	56.3
Monterey–Salinas	67.8	Butte	56.3
Beckley–Bluefield–Oak Hill	67.7	San Diego	55.8
Odessa–Midland–		Champaign & Springfield–	
Monahans	67.3	Decatur	54.8
Lima	67.1	Syracuse	54.7
Yuma–El Centro	65.8	Charleston–Huntington	54.3
Roswell	65.8	Waco–Temple	54.1
Bakersfield	65.2	Topeka	53.7
Cheyenne–Scottsbluff–		Presque Isle	53.4
Sterling	63.7	Chico–Redding	53.2
Zanesville	63.1	Reno	53.1
Wilkes Barre–Scranton	63.0	Tyler	52.9
Binghamton	63.0	Grand Junction–Montrose	52.9
Wheeling–Steubenville	62.9	Medford–Klamath Falls	51.9
Eureka	62.6	West Palm Beach–Ft. Pierce	51.8
Abilene–Sweetwater	62.4	Macon	51.5
Utica	61.9	Twin Falls	51.4
Lafayette, In.	61.2	Alpena	50.3
Greenwood	60.7	Mankato	50.3

Source: Nielsen (February 1982)

television, selection has largely been on the basis of age and sex. Only in print has there been a real effort to zero-in on highly selective audiences and environments through special interest magazines.

This relative "mass nature of the mass media" has often frustrated advertisers seeking to better concentrate their marketing dollars against selective marketing targets. And it has been even more frustrating to the smaller advertisers of non-mass-market products who, to paraphrase John Wanamaker, often feel that "at least half of their advertising dollars are wasted but aren't sure which half!"

Targeting Audiences and Messages

Cable video allows the mass-market advertiser to target more closely on specific marketing segments that he regards as having above-average consumption potential. And for the small budget, non-mass-market advertisers, cable can provide a very precisely targeted video environment he might be unable to afford in network or spot television and has had to seek out in print. When this more highly targeted video environment is coupled with the use of a more informationally oriented, in-depth advertising message, the results can be an overall increase in advertising effectiveness. (See Figure 6-1.)

Following are 11 examples of the use of different cable avenues for different advertisers, recognizing that "all are not created equal" but have different objectives in terms of audience selectivity, program compatibility, and geographic targeting.

Cable Health Network

The Cable Health Network delivers information on health and physical well-being 24-hours a day. An insurance company could sponsor a show on nutrition, or a pharmaceutical company could sponsor an exercise show.

American Baby

The first cable program series devoted to educating the new and expectant mother, the "American Baby" series (Satellite Program Network) provided information and live demonstrations of all phases of childbirth and early life. In addition to its cable exposure, the series is distributed free to childbirth classes throughout the country.

Medical World News

Based upon "Medical World News," the newsmagazine for doctors, this series airs on the Satellite Program Network. It includes segments dealing with everything from fitness to forensic medicine, "Your Family Doctor," and various controversial health-related issues. The environment provides a means of reaching a dual audience—physicians and people concerned with issues involving their physical well-being.

Family Guide to Boating Fun

This newspaper supplement represented a new vehicle for boating advertising and combined manufacturers' national ads with local dealer tie-ins. It was coupled with a point-of-purchase kit and contest drawing and a half-hour cable show (Modern Satellite Network) featuring products of the advertisers. It was a classic case of in-depth information directed to a concentrated (boat-owning) target audience.

Mitsubishi and the Mirage Bowl

The 1981 Mirage Bowl (ESPN), which took place in Tokyo and featured Western Conference teams, San Diego State against the U.S. Air Force Academy, was sponsored by Mitsubishi. This is a good example of how Mitsubishi, a Japanese company, dominated a specific event on cable and provided advertising and promotion support for its products in a foreign market.

Military Cable TV Network

The Military Cable TV Network is aimed at the active and retired military market and sells time on local cable systems that reach 54

FIGURE 6–1. Media Communications Segments

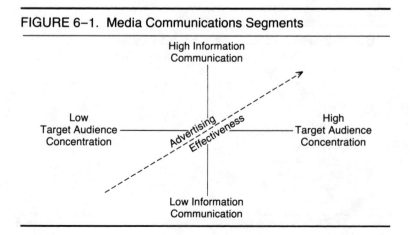

military bases around the country. While the cable systems reach more than just the military audience (e.g., San Diego delivers a greater audience than the naval base alone), the concentration of military personnel is greater than would be offered by broadcast stations.

CBS Cable and the Affluent Reader

For the advertiser seeking to reach a very sharply upscale target, magazines have been the primary medium. To document their ability to reach this audience with the market segmentation of magazines and video impact of television, CBS Cable ordered a special analysis by the Simmons Market Research Bureau. Simmons analyzed the magazine-reading habits of households that reported viewing cultural and performing arts television programming *and* were located within the coverage area of cable systems carrying CBS Cable. Compared to the national average, the CBS Cable household was more likely to read the magazines listed in Table 6-5 regularly by significant amounts.

H & R Block and Local Cable

An H & R Block district manager conducted a Tax-A-Thon on his local cable system for 24 hours on the weekend immediately preceding the deadline for filing. In addition, he used the cable system in a variety of other ways:

- Tax tips were aired weekly on a community activities program.
- An annual H & R Block Banquet was taped and a 30-minute show was cablecast.

TABLE 6–5. Readership of Upscale Magazines by Potential CBS Cable Households Compared to National Average

Business Week	+38%
Forbes	+32
Fortune	+87
Scientific American	+45
U. S. News	+49
Wall Street Journal	+43
Any Epicurean Publication	+49
Omni	+84
New York	+83

Source: SMRB Special Analysis (November 1981)

- Filmstrips of a new H & R Block office were shown.
- A live interview show was held.
- H & R Block executives were presented on cable.
- Tax-related items were carried on the weather scan.

"Woman's Day USA"

A video translation of the monthly magazine "Woman's Day USA" is produced by Young & Rubicam for its client General Foods and carried on the USA Network. Featuring a weekly menu planner, a shopping guide, and cooking features, it runs several times a week on Wednesdays through Saturdays (peak shopping days). The program does not, however, stand by itself. It is coupled with a number of two-minute "newsbreak" segments, "Today's Meal," that are scattered throughout the daily USA schedule. These provide instruction on how to prepare the key elements of the meals featured on the half-hour show. The half hour encourages viewing of the two-minute segments, and the two-minute segments enhance the value of the half hour. It is a fine example of new media synergy where one vehicle reinforces the value of another.

Rodale Press—Print and Video Synergy

Rodale Press publishes a series of magazines aimed at helping people improve the quality of their food, health, homes, and lives. These include *Prevention, Organic Gardening, Bicycling, Rodale's New Shelter,* and *The New Farm.* Rodale brought the subject matter of these publications to cable in Rodale's "Home Dynamics," a 13-part series on the CBN Satellite Network. It focused on how to produce your own low-cost food and energy at home, thus allowing advertisers to reach this audience both via magazines and video.

To provide an even further extension of this multimedia approach to advertising, most of the programs in the "Home Dynamics" series are available as home video cassettes. Sponsors of the cable series have the opportunity of placing their advertising in these cassettes. These are then offered to groups like gardening clubs, solar energy societies, retirees' associations, etc.

Quaker Oats and the Weather Channel

The Weather Channel, owned by "Good Morning America" meteorologist John Coleman and Landmark Communications, provides

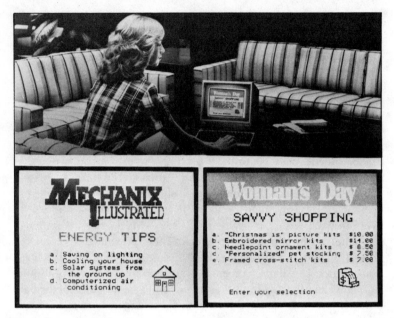

The "Woman's Day" test of videotex showed a high degree of consumer interest in shop-at-home services.

national, regional, and local forecasts, along with weather-oriented features.

As a charter national advertiser on The Weather Channel, Quaker Oats sponsors special "cold weather reports" to promote its hot cereal products when the temperature dips below 40 degrees in a large number of markets. In the past, advertisers have developed weather associated features to run on radio when the temperature hit a certain level, when it rained, or when snow was being forecast. The Weather Channel allows this audio advertising to become video oriented.

Two Approaches to Cable: Media Audience and Response

The Media Audience Approach

Media are traditionally evaluated on the basis of the number of potential prospects for a product or service an advertiser can reasonably expect to be in the audience of the programs or publications where the advertising appears. For television this generally means talking about the millions of impressions or the reach and frequency delivered against viewers of various demographic characteristics. From a pure media standpoint, cable can provide advertisers with low-cost continuity of exposure they may not be able to afford on broadcast television. An advertiser might find that the elimination of only one commercial in a television schedule will allow him to purchase an extended cable flight aimed at his target audience. For example, for the $150,000 it might cost for one network commercial on ABC's Monday Night Football, an advertiser could purchase a 10-week, fourth-quarter 1982 schedule of 7 spots per week on ESPN, the 24-hour-a-day sports network.

All evidence points to the continued growth of cable and the increasing channel capacity that will provide many viewers with 50 or more options to choose from. (Table 6-6 gives a 50-channel cable prototype.) And this increasing channel capacity, coupled with the expansion of cable and other over-the-air new media into new markets, will result in a year-to-year decline in broadcast audience ratings. For example, the situation with prime-time network television might look something like that shown in Table 6-7.

As would be expected, the impact of the new media will be the greatest in those homes that have the most of it. Today, this means in pay cable homes. It also will be greatest during the warm weather network rerun season (see Table 6-8). And, finally, across the day

---

TABLE 6–6. A 50-Channel Cable Prototype

3–Local Net Affiliates	3–Ethnic
3–Indies	3–Religious
2–News	2–Children's
2–Sports	3–Education and Information
3–Imported Stations	4–Local Access
4–Movies	2–Music
2–PBS	1–"How To"
4–Entertainment and Culture	1–Women's
1–Weather	1–Games and Participation
1–Health	1–Info Retrieval
1–Senior Citizens	1–Business and Finance
2–Shopping and Infomercial	

there are significant differences, with network ratings suffering the most in pay cable homes in prime-time and early fringe hours (Table 6-9).

The full impact of cable on broadcast audience levels is even more evident when one examines individual markets with different levels of cable penetration. The result is that an advertiser buying a television schedule, whether nationally or locally, can expect to find substantially lower ratings delivered in cable households than in non-cable households. While the differences will vary by market and by time of day, Table 6-10 shows what the situation might be in a time and in a place where 30 percent of the homes have cable.

TABLE 6–7. A 10-Year Forecast of New Media Impact
on Network Audiences (Percent)

	Cable And New Over-Air Penetration*	Primetime Network TV Audience Estimates		
		Homes Using Video	3-Network Shares	Average Network Rating
1981	30	59.5	82	16.3
1982	34	60.0	79	15.8
1983	38	60.5	77	15.5
1984	42	61.0	74	15.0
1985	46	61.5	72	14.8
1986	50	62.0	70	14.5
1987	53	62.5	68	14.2
1988	57	63.0	65	13.7
1989	60	63.5	63	13.3
1990	63	64.0	61	13.0

*Note: includes STV, MDS, and DBS

TABLE 6–8. 1981 Prime-time Network Ratings in Cable
and Noncable Homes (Percent)

	Winter (February)	Spring (May)	Summer (July)	Fall (November)
Noncable Homes	18.6	16.2	13.7	17.9
Basic Cable Homes	18.4	14.8	11.8	16.3
Pay Cable Homes	16.6	13.1	9.9	15.3

Source: Nielsen, *Cable TV Status Report*

The differences in Table 6-10 are seen not only in the ratings but in the reach and frequency of a specific broadcast schedule. For example, if an advertiser in the past looked only at the delivery of his plan among all homes in a market or in the country (including both cable and noncable), the numbers might look like those in Table 6-11. Among cable homes, however, the total ratings, the reach, and the frequency all fall below the level in Table 6-11. (See Table 6-12.)

To compensate for this underdelivery in cable homes, an advertiser can buy spots either locally or nationally on the cable satellite networks. Ted Bates & Co., one of the nation's largest advertising agencies, in fact recommended to its clients that 5 percent of their television network prime-time budgets be transferred to Superstation WTBS as a way of recouping audiences that the networks had lost to pay cable homes. The recommendation followed the analysis by Bates of a number of special Nielsen studies that showed a significant underdelivery of network schedules in pay cable households. WTBS was

---

TABLE 6–9.   1981 Cable and Noncable Home Network Ratings for Different Dayparts

	Primetime (Percent)	Daytime (Percent)	Early Fringe (Percent)	Late Fringe (Percent)
Noncable Homes	16.6	7.1	11.2	7.0
Basic Cable Homes	15.4	6.5	11.0	5.9
Pay Cable Homes	13.7	6.8	8.3	7.2
All Homes	15.9	6.9	10.7	6.8

Source: Nielsen, *Cable TV Status Report* (February, May, July, November)

TABLE 6–10.   Possible Ratings for Cable and Noncable Homes (Percent)

	All TV Homes (100%)	Noncable Homes (70%)	Cable Homes (30%)
Primetime Homes Using TV	60	59	62
Local Shares	85	95	70
Local Ratings Available	51	56	43
Average Network Rating	17	19	14

Source: Arbitron Special Report, "The Impact of Cable on Spot Television Buying," 1981.

TABLE 6–11. Past Four-Week Reach and Frequency

	All TV Homes (100%)
GRPs a Week	120
Reach	80%
Average Frequency	6.0
Total GRPs	480

Source: Arbitron Special Report, 1981.

specifically recommended because it offered the largest cable audience on a national basis. Advertisers interested in reaching more specialized cable audiences, however, might also use other channels of a more special interest nature to help fill this deficiency.

The Response Approach

Many marketers recognize the importance of advertising response rather than simple audience contact. They do not necessarily evaluate a media buy on the basis of the cost per thousand people it reaches. Rather, they are more concerned with the cost per response generated.

An example of the use of this approach in evaluating a cable opportunity could involve a local auto dealer who has an opportunity to do a 15-minute program for $100 on the subject of "How to Buy a New Car." If 10,000 homes subscribed to the cable system and only 50 people watched the auto dealer, most media planners would consider the buy both ineffective and inefficient: ineffective because it only reached 50 people out of 10,000 homes (less than a 1 percent rating), inefficient because, at a cost of $100, its cost per thousand was a whopping $2,000!

The astute auto dealer would, however, have looked at this as an outstanding advertising response generator. First, he got his message across to a full 50 people, far more than came into the showroom on a single day. And, second, it only cost $2 to talk to each of these people for 15 minutes. He would probably have been willing to pay two or three times this amount to anyone who came into his showroom and listened to him discuss how to buy a new car for a quarter of an hour.

In another case, a pet food manufacturer might seek a unique environment in which to tell the story about his new, premium-priced dog food. He has the opportunity to sponsor a prestigious dog show on

TABLE 6–12. Present Four-Week Reach and Frequency

	All TV Homes (100%)	Noncable Homes (70%)	Cable Homes (30%)
GRPs a Week	120	130	97
Reach	80%	84%	72%
Average Frequency	6.0	6.2	5.4
Total GRPs	480	520	386

Source: Arbitron Special Report, 1981.

cable, but on examining the potential audience delivery, he finds that it will cost at least $20 per thousand women reached—four times that of his current television plan. Is he discouraged? Not necessarily. First, he knows that the show provides a far more compatible environment for the product than his normal network or spot television programming. And, second, he feels confident that he is reaching a responsive audience and that the majority of the viewers will own a dog and represent good prospects for a premium dog food. As a matter of fact, if he were to zero-in on his basic late-fringe television buy, and go beyond CPM women to CPM dog-owning women in an age group and income group that might be regarded as true prospects, his dog show cable opportunity looks like a real bargain. (See Figure 6-2.)

Cable and the Direct Marketer

For years it's been recognized that no one knows how to motivate sales more or measure the impact of advertising on sales more than does the direct marketer. The national advertiser, for example, generally must evaluate media effectiveness on the basis of the size of the audience reached relative to cost per thousand, consumer awareness studies, and product share trends. The direct marketer, however, can bypass the normal media measurement services and evaluate the effectiveness of individual vehicles on the basis of the sales or sales leads they produce. Over a period of time the direct marketer sees what works and what doesn't work and eliminates the ineffective vehicles from the media mix. (It's always been a bit amusing that direct marketers generally regard the television time periods that national advertisers feel are least attractive as extremely profitable in terms of pay-out per dollar expended—namely, late night and Sunday morning.)

FIGURE 6–2. A Late-Fringe Television Buy vs. A Dog Show

$5 a 1000		All Women		
$10 a 1000	25–54		vs.	$20 per 1000 Women
$25 a 1000	$30,000 Income			
$50 a 1000	Who Own a Dog			

Direct response advertisers traditionally have concentrated on the use of direct mail, print with coupons or inserts, and television. The targeted appeals of the new electronic media coupled with their extended message length flexibility have made them extremely attractive to the direct marketing industry. And just as the direct mail order business is growing at a rate nearly twice that of retail, it would not be surprising to see its use of cable increase equally fast.

The Appeal of Cable to Direct Marketers

The segmented appeals of the special interest cable networks can be an effective direct line of sales for direct mail marketers. Perhaps one day there will even be "direct video lists," just as there are direct mail lists today. Fanciful? Mail order financial and insurance companies already can use business news channels to get new clients. Direct-marketed kitchen and houseware products find receptive audiences among the how-to and daytime women's service shows, and physical fitness products can utilize the cable health and medical networks.

The 800 telephone numbers, which today allow viewers to purchase the products they see on television or cable, may be the predecessor of the pushbutton interactive technology that will be developing more fully in the second half of this decade.

Direct response marketers will find the segmented appeals of cable coupled with the attractive ability to use nonstandard commercial lengths. In short, they will have the time required to convey the information they need to convey.

"The Great Catalogue Guide": A Unique Cable Test

In late 1980, the Direct Mail Marketing Association sought an innovative and dramatic communications program directed to consumers to achieve the following objectives:

1. Communicate the benefits of shopping by mail, e.g. convenience, value, selection, etc.
2. Devise a convenient and interesting way for consumers to identify and contact individual mail order catalog firms offering the products they want.
3. Generate 2,000 requests for this information.
4. Test cable's ability to achieve this within a reasonable budget.

A 120-second direct response commercial offered, through an 800 toll-free number, the DMMA "Great Catalogue Guide" as a free premium. The "Great Catalogue Guide" listed more than 550 consumer catalogs in 27 product categories. Postcards for consumers to fill out to request catalogs they would like to receive were also included in the "Guide." The commercial was transmitted by satellite to 600 local cable systems, which aired it from October 20 to November 3, 1980. The total budget allocated to the program (the commercial, the "Catalogue Guide," fulfillment, and media) was approximately $22,000.

More than 7,000 consumers called during the two-week period the commercial was aired to request the "Catalogue Guide," a 250 percent increase over the goal of 2,000 requests. This was a cost per inquiry of about $3. In addition to the excellent response, reaction from DMMA membership was overwhelmingly positive and supportive.

One Marketer's Opinion

Alvin Eichoff, President of A. Eichoff and Co. (now a part of Ogilvy & Mather), summed up the value of cable by saying:

> "Broadcasting's 30 and 60 second commercials are artificial. Commercials should be able to be as long as they need to be and, as cable outlets develop, significantly longer commercial spots will become available. Many ad execs see such developments working to make it possible to sell such big-ticket items as automobiles via televised direct response ads. You'll be able to take the time to show the viewer the car's engine, explain why rack and pinion steering is the best, and tell the viewer what is so good about independent suspension. That would be a lot better than just showing a big cat jumping on a car!" (*Advertising Age*, November 16, 1981, p. S-14.)

Cable and Direct Mail in Tandem

In combination, cable and direct mail can be an extremely powerful media package. For example, a 12-part cooking course can be created for cable at a very modest cost, perhaps even using the home economics department at a university. The course could be directly tied-in with a direct mail program offering the recipes, product coupons, a cookbook, and even "Videocooking Tapes." On a local basis, this could even be tied in with special grocery store promotions. The concept need not be confined to cooking but could be extended to how-to courses on car care, on fashion and grooming, or on health maintenance.

Sports Shack: A Success Story

One of the most successful direct response cable campaigns involved the Martin Lambert Advertising Agency and its client, Sports Shack. Sports Shack was a company involved in selling sporting goods store franchises. Their advertising was confined largely to newspapers and business publications. The company avoided using television because of its absolute cost and because they felt that the number of persons interested in and able to afford a $36,000 franchise was minute.

Sports Shack agreed, however, to test a flight of 15 60-second spots on the Entertainment and Sports Programming Network. If as many as 30 to 45 leads had been secured (2-3 per commercial), the campaign would have been regarded as a success. By the time the flight ended, over 600 people had called for further information. Of these, 86 requested interviews with the Sports Shack field sales staff. And, of these, 40 prospects agreed to fly to Minneapolis at their own expense to finalize contracts. For Sports Shack, cable proved to be an excellent response medium.

Information in the Home

It will be a long time before it is determined just how successful the new media can be in encouraging people to shop from their living room chairs. Some people will like the idea, others won't. And, for some products, it will be successful, while for others it will probably be disastrous. Aside from actual product purchases, however, the television set can be used to distribute many forms of consumer information.

Advertising and Videotex

For advertisers interested in providing consumer information and transaction services, videotex offers a number of opportunities. (See Table 6-13.) Already videotex experiments have been conducted by CBS, AT&T, Knight-Ridder Newspapers, Dow-Jones and Co., Reader's Digest, Cox Cable, and the Times-Mirror Company.

In Ridgewood, New Jersey, CBS and AT&T joined together in the fall of 1982 to test for seven months the ability of videotex to achieve consumer acceptance in a number of different areas including:

- To call up information such as news, sports, weather, financial data, encyclopedias, almanacs, schedules, etc.
- To conduct transactions such as banking, bill paying, shopping, making travel and entertainment reservations.
- To participate in computer-based education.
- To be entertained by computer-based games, quizzes, and features.
- To communicate electronically with other users—a technique known as electronic mail.

This is but one of a number of experiments with computer-based home information systems that have been and are going on around the country.

From an advertising standpoint, marketers can use videotex at any one of three different levels.

Level One—To Run an Advertising Message. A variety of advertisements of differing sizes can be developed and run in product- or service-related frames of information. Table 6-14 gives some examples.

On Labor Day 1981, Field Enterprises in Chicago began telecasting its "Night Owl" teletext service over WFLD-TV, daily from midnight to 6:00 A.M. For the person who had been out in the evening and hadn't had a chance to see the paper or look at an evening newscast, "Night Owl" offered a continually updated 20-minute review of the news, sports, and weather, together with a variety of leisure and entertainment programming. From a creative advertising standpoint, stores and products who used Field's "Chicago Sun-Times" could, for a very modest additional charge, use "Night Owl" to highlight and supplement their next day's ads. For example, an appliance store

TABLE 6–13. U.S. Videotex Tests and Services

Test/Service	Location	Date	Transmission Method	Number of Subscribers	Services
Alternate Media Center/ WETA (Telidon) *TEST*	Washington DC	1981-1982	Broadcast one-way	50	*Washington Post, New York Daily News,* US Weather Bureau
Belo BISON *SERVICE*	Dallas TX	1982	Phone lines	150	*Dallas Morning News,* UPI, Reuters, local entertainment guides
CBS/AT&T *TEST*	Ridgewood NJ	1982	Phone lines two-way	200	News, weather, sports, home shopping, banking, CBS Publications
CBS Extrovision (Antiope) *TEST*	Los Angeles CA	1981-1982	Broadcast one-way	100	Sporting goods, travel packages, airline schedules
Chase Manhattan *TEST*	New York City area	1982	Phone lines two-way	—	Home banking
Chemical Bank/ Pronto *TEST*	New York City area	1982	Phone lines two-way	200	Home banking
Citibank/Homebase *TEST*	New York City area	1982	Phone lines two-way	100	Home banking
CompuServe *SERVICE*	National	1981-1982	Phone lines two-way	19,000	Games, news, 11 local newspapers, Comp-U-Card (shopping service)
Cox/Indax/HomeServ *TEST*	San Diego CA	1982	Cable two-way	300	Shopping, banking, The Source
Dow Jones *SERVICE*	National	1981-1982	Phone lines two-way	30,000	Stock quotes, *Wall Street Journal, Barron's,* financial services Media General and Disclosure On-Line, and money market services

TABLE 6–13. (continued)

Dow Jones/Sammons *SERVICE*	Park Cities TX	1981-1982	Cable two-way	35	*Dallas Morning News*, AP newswire, entertainment and restaurant guides, sports, airline schedules, *Wall Street Journal*
Field Electronic Publishing-Keyfax (Prestel/Ceefax) *SERVICE*	Chicago IL	1981-1982	Broadcast one-way	110	*Chicago Sun-Times*, news, weather
Knight/Ridder Viewtron/AT&T *TEST*	Coral Gables FL Southern FL	1980-81 1983	Phone lines two-way	150	*Miami Herald*, Dow Jones, Sears, J. C. Penney, B. Dalton Booksellers, Eastern Airlines, *Consumer Reports*, Grand Union
The Source *SERVICE*	National	1981-1982	Phone lines two-way	14,500	Electronic mail, UPI news, stock market quotes, travel information, restaurant guides, discount buying services
Time, Inc. *TEST*	San Diego CA Orlando FL	1982	Satellite/cable one-way	150 150	Sports, arts, entertainment, self-help, travel, transportation
Times-Mirror (Telidon) *TEST*	Los Angeles area	1982	Two-way cable Phone Lines	150 200	Teleshopping, banking, airline schedules, entertainment reservations, Comp-U-Card, *Los Angeles Times*, magazines
Warner Amex Qube *SERVICE*	Columbus OH Cincinnati OH Pittsburgh PA	1977-1982	Cable two-way	34,000 — —	Games, news, weather, electronic mail, CompuServe

Source: *Videography*, April 1982.

featuring specials on a dozen or more items was able to use individual "Night Owl" textual frames to provide information on each individual item.

Level Two—To Seek Direct Response. Products or services that have the potential of being ordered via the terminal, through an attached telephone handset, or through couponing in conjunction with a local retail outlet can use this avenue. While today's leading direct marketers (see Table 6-15) may be expected to become leaders in the use of videotex for direct response, other advertisers can find it of value simply to offer free samples, coupons, or product information.

Level Three—To Supply Program Data. In this case the advertiser would supply not just advertising, but a full complement of informational programming. For example, a pet food company could provide the basic data on the care of pets. A diversified food company could provide information on a variety of recipes, or an automobile manufacturer could provide information on safe driving and car maintenance. The Shell Answer Man, for instance, could easily become a regular videotex feature.

Earlier in this book, I noted how the new media would impact not just on television, but on all media, including newspapers. Because of this, newspapers are experimenting with ways of effectively using cable and data transmission services to their advantage.

One example of such a joint effort is an Associated Press/CompuServe classified advertising videotex experiment. In 1982 the Associated Press, in conjunction with CompuServe, Inc., and 11 major newspapers, embarked on a test of a national network of a two-way interactive home information system in which subscribers could re-

---

TABLE 6–14. Possible Vehicles for Different Products and Services

This Product or Service	Would Run In
Automobile	Travel Tips
Beer	Sports Scoreboard
Camera	Community Calendar
Financial Institution	Money Management Tips
Food Product	Menu Suggestions
Insurance Company	Medical Tips
Pet Food	Lost and Found Pets
Sporting Goods	Golf or Tennis Tips

trieve employment, real estate, and automotive classified ads. CompuServe is a computing services company that operates a nationwide interactive home information or videotex service for approximately 20,000 home subscribers.

TABLE 6–15. Sales Volume of Leading Direct Marketers, 1981 (estimated figures)

Company	Sales volume (in millions)
Sears, Roebuck & Company	$1,646
J. C. Penney Company	1,537
Montgomery Ward & Company (Mobil)	1,231
Time-Life Books	400
Spiegel (Otto Versand)	387
Franklin Mint Corporation	375
Fingerhut Corporation (American Can Company)	300
Aldens (Wickes Corporation)	260
Signature Financial Marketing (Wards)	200
Columbia Record Club	195
Publishers Clearing House	170
American Express Company	140
Grolier	135
L. L. Bean	132
Heath Company (Zenith Corporation)	115
Avon Fashions	100
RCA Music Service	100
Herrschner, Brookstone & Joseph Bank (Quaker Oats)	100
Hanover House Industries	100
Spencer Gifts (MCA)	85
Current Incorporated (Looart)	70
Dreyfus Corporation	60
Book-of-the-Month Club (Time Incorporated)	60
Swiss Colony	53
Roaman's	50
Eddie Bauer (General Mills)	43
Miles Kimball	40
Lillian Vernon Corporation	40
Horchow	40
Lee Wards (General Mills)	35
Figi's (American Can Company)	35
Harry & David	35

Source: Direct Marketing Association

From Information to Shopping

The longest running shopping service is the Modern Satellite Network's "Home Shopping Show," offering 8½-minute and 30-minute segments of in-depth information about a product or service. The show runs five days a week.

This has been joined by a number of other information and shopping ventures either in operation or in the testing stage. Included are:

- Shopping by Satellite—a joint-venture shop-at-home service between Comp-U-Card and Metromedia airing two hours a day on the CBN Satellite Network.

- Narrowcast Marketing USA's "Showcase"—a service composed of 5½-minute (or longer) infomercials for products in the consumer electronics field.

- "The CABLESHOP"—an experiment by J. Walter Thompson USA and Adams-Russell Company. Viewers use their telephone to call up informational messages they want to view on their television screens.

- "The Shopping Game" on the Satellite Program Network—home viewers order via a toll-free 800 number the same prizes that contestants compete for on the show itself.

- "The Sharper Image Living Catalog" on the Satellite Program Network—a 30-minute video version of the San Francisco-based Sharper Image catalog. Nine to 12 Sharper Image products are demonstrated and sold on each show.

- "Easy Shopping" on the UTV Network. A wide range of products, from housewares to high fashion, are presented, discussed, and sold via a toll-free 800 number. The *entire* UTV Network is devoted to viewer participation from the home.

Comp-U-Card also developed one of the more sophisticated computerized view-and-shop-at-home systems. By using a keyboard to input product questions, the consumer secures facts and details on a wide-range of products. After providing the specifications, a central computer will offer a selection of makes and models that meet the criteria. The consumer can then order the product and arrange for its shipment to the location.

Some products obviously respond better to the Comp-U-Card shopping system than others do. Since it currently lacks a still or moving picture capacity (with information consisting of text and computer-animated pictures), its greatest potential currently involves those products for which viewers don't need to see a full life-like reproduction.

There is no question but that a major weakness of any of the videotex services for home shopping is that many products cannot be adequately shown by computer graphics. The International Interact Corporation is working to get around this problem by providing a magazine-quality service to advertisers, information providers, and cable companies. It would provide full-color still photographs (with a voice-over) on a simple (nontechnical) terminal in the home. Products would be displayed:

- By store or company, as is seen in individual catalogs today.
- By individual item, perhaps showing all products in a specific category (television sets, tires, and toasters) from all the retailers participating in the system.
- By individual price range (television sets under $100—$100 to $300—$300 to $500—over $500).
- In a "Space Store," where shoppers would literally take a "video walk" through the store and be able to stop in individual departments for detailed looks at the products.

Interact's hope is to create more of a feel of shopping than exists in the present computer home shopping services.

Your Imagination: The Only Limitation

More and more advertisers are experimenting with ways formerly reserved exclusively for print. Over the years, people have grown used to getting their news and information from the television set, and businesses are trying to capitalize on this.

Real estate firms are experimenting with the use of cassettes to "list" and "show" homes. The objective is to let prospective buyers screen out those homes they have no real interest in and save driving time for those houses that are real possibilities.

The first video cassette news release was taped last year at a 45-minute press conference in New York announcing Pratt & Whitney's sale of a $600 million commercial aircraft engine. Within three hours of the news conference, the tape was edited and released by satellite to cable systems across the country.

Also last year, International Paper Company came out with the first video annual report. Its financial data and a video tour of its operations were highlighted in a half-hour cassette made available to stockholders in all tape sizes and aired on cable systems across the country.

A video trade show went one step further. The first such one was the "National Consumer Electronics Showcase" developed by Narrowcast Marketing. Such products as video disc systems, cassette recorders, electronic games, and home computers were demonstrated in 5- to 11-minute infomercials transmitted by satellite to approximately five million homes across the country. The messages were incorporated in an editorial environment that included updates on general product developments in personal communications systems, electronic games, advanced TV, home security, and similar items. The showcase provided a national video supplement to the regular live trade shows where advertisers regularly display and demonstrate their products.

The point here is that opportunities for the use of video to transmit informational advertising on all subjects, products, and services are basically limitless.

CableCash, an advertising experiment by the Cable Coupon Network, will roll out in January 1983.

As a matter of fact, new media advertising can extend all the way to the bills that cable subscribers regularly receive for service. One thing that all cable subscribers have in common is the monthly bill they receive from their cable system. CableCash, a 32-page coupon book delivered to subscribers with their monthly statements, attempts to capitalize on this vehicle. Beginning in August 1982, the Cable Coupon Network began a roll-out of CableCash to a total January 1983 projected national circulation of 16 million. In addition to seven regional breakdowns, individual markets are available for more sharply pinpointed delivery. For companies currently using coupons and/or cable as part of their advertising effort, this can be an interesting addition to the marketing plan.

In a nutshell, today's advertiser is really limited only by his or her new media imagination.

CHAPTER 7

Creating and Producing Advertising for the New Media

When Marshall McLuhan wrote that "the medium is the message," he was obviously thinking only of television and not of the vast potential for advertising in the new media. Had McLuhan done so, he might more properly have said, "the medium is the message, *except* in cable, where the message can be the medium!"

Throughout this book, I have identified 10 major values of cable that must be considered in the advertising planning process:

1. Highly specialized programming allowing advertisers to zero-in on highly selective target audiences.

2. Low unit costs per announcement.

3. Ability to build a high frequency of exposure.

4. Flexibility of advertising message lengths and forms.

5. Sponsorship opportunity with program identity.

6. Product exclusivity in programs.

7. Ability to test at low media costs.

8. Opportunities to tag advertising messages with direct response offers.

9. Ability to localize advertising for franchisees, dealer organizations, and wholesale and retail sales forces.

10. Creative commercial opportunities that match viewer life-styles with program environment.

Concentrating on Message Content

Until recently, most advertiser involvement in cable has concentrated on the medium and various approaches for using it to deliver existing broadcast commercials. It is equally important to focus on the advertising message that is delivered. In other words, cable's values must be brought alive through advertising creativity as well as through creative selection of the cable media vehicles.

The cable viewer tends to be upscale in education, occupation, and income. This viewer is a seeker rather than a passive receiver of information and entertainment and is attracted to the program content of cable in many of the same ways that a reader is attracted to the targeted editorial content of a magazine. As a result, there probably is less need for the advertiser to concentrate on attention-getting commercial devices than there is in broadcast television. The consumer will seek targeted cable channels of communication, so the advertiser is able to concentrate more on message content than on viewer contact. (See Figure 7–1.)

For example, a typical 30-second automotive commercial relies on very strong attention-getting devices to attract the viewer. There is a relatively short amount of time in which to deliver an in-depth message. To demonstrate the potential of cable to deliver a high-information-content message, a two-minute "racer testimonial" developed by J. Walter Thompson USA stressed the selection of a Ford Mustang as the best car to race in the International Motor Sports Association American sedan class. "Inside Ford Racing" emphasized the merits of the suspension system and application of small-engine technology that pointed to a new breed of fuel-efficient race cars. Rob McFarlin, champion driver in the IMSA import sedan class, outlined how today's advanced Ford technology helped build better automobiles for the American consumer. The commercial, produced on tape at Road Atlanta, was designed to be placed in auto racing on cable. During the events, Rob was racing, giving the "testimonial" even greater impact. Since viewers have a natural interest in racing, the attention of the advertising message can be focused more on message content than on attracting viewer contact. In this case, there was a creative commercial opportunity to match up viewer lifestyles with program environment.

FIGURE 7–1. TV and Cable Communication

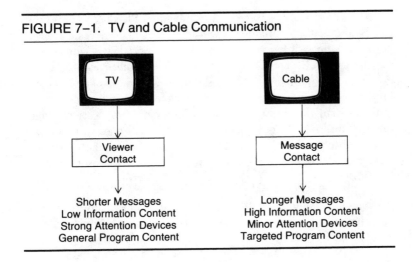

The Cable Creative Paradox: More Messages for Fewer Dollars

All this does not mean, however, that cable advertising can be "dull." Creativity is still vital, since viewers will always react far better to a well-executed 30-second commercial than to a poorly executed two-minute message. What will change is the approach of the message and the devices employed to communicate. The emphasis will be more on maintaining interest throughout the length of the message than on attracting initial attention to it. As a result, creative directors and producers will need to develop visual techniques that allow for the attractive and effective presentation of large quantities of information.

And, of course, they will have to do this at budgets substantially lower than for today's broadcast commercials. A $50,000, $100,000, or $500,000 commercial makes sense if the media budgets behind them are $1,000,000, $5,000,000 or $10,000,000. In cable, however, an advertiser will generally be running a large number of messages in one or more very low-rated programs with very low media costs. These cable messages will undoubtedly have a much shorter effective life than a broadcast commercial does since the reach of any individual message will be low and the corresponding wearout factor will be relatively high.

Wearout will especially be a problem for infomercials—in-depth cable presentations of information. In broadcast television, media planners talk about the "effective reach" of a message and the need for a minimum of perhaps three exposures before it has been communicated sufficiently to produce a viewer impact. The upscale viewer, however, who is interested in a product or service and who seeks a specific channel, may actually be adequately reached with only one exposure of an infomercial. Similarly, the overexposure of an in-depth, long-form cable message may have a far more negative effect on the prospect than the overexposure of a 30-second broadcast message. (See Figure 7–2.) In other words, marketers will need to produce a greater number of cable advertising elements for substantially fewer dollars than they are accustomed to producing for broadcast television.

The Need for New Production Techniques

To meet the challenge of the Cable Creative Paradox, advertisers and agencies will be called upon to develop entirely new production techniques for the medium.

In 1982, advertisers will invest approximately $6 billion in network television, about 10 percent ($600 million) of which will be spent shooting commercials. The average broadcast 30-second commercial cost $55,000 to produce in 1982. In cable, however, creative directors and commercial producers must learn to develop messages of several minutes in length and bring them in at budgets of $5,000, $3,000, and, in some instances, under $1,000.

For the advertising agency, cable will undoubtedly result in new financial strains and needs for new compensation plans. Traditionally, agencies have been paid on the basis of a 15 percent commission on media purchased. Generally, they have fought the concept of flat

---

FIGURE 7–2. The Cable Paradox

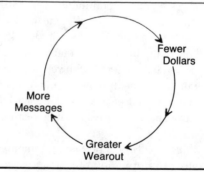

fee compensation by clients for fear that fixed fees might not adequately cover variable costs.

Now, however, cable is totally changing the complexion of compensation. The small dollar amounts involved in early cable placements make it impossible for agencies to profit if they must develop special commercials via the commission route. Agencies will find that fees become a necessity if they are to be able to afford the time to produce effective commercials and campaigns for cable. As Paul Kagan, a specialist in the field of broadcast and cable consulting, pointed out: "Cable may prove to be an unexpected boon to ad agency profit and loss statements, not because of the springing loose of cable ad budgets, but due to the fall-out of fees after the cable detonation."

The New Media Message That Motivates

It must be recognized that the goals of cable communications and broadcast commercials are the same: the sale of a product or service to a prospective customer. One secret in the development of effective cable advertising approaches is to focus the maximum amount of creative attention on the *message content*. The three important terms in creating an effective cable commercial are the *media message that motivates* (i.e., sells), the *discernible product attributes*, and *significant consumer desires*. The discernible product attributes are those facets of a product's construction, appearance, and uses that will satisfy significant consumer desires—all of those reasons why people will want the product and buy it.

In the development of a 30-second commercial, time limitations dictate that all attention must usually be focused on a single discernible product attribute that will satisfy a single significant consumer desire. There is obviously not much time to accomplish this, especially when a portion of the message—often a substantial portion—must be directed at simply gaining and holding the viewer's attention.

Cable, however, offers the luxury of time—time to develop a *message that motivates*—during which an advertiser can show and discuss *many* attributes of the product (or service) that may satisfy *several* consumer desires. Just as in the case of the broadcast commercial, the message must be interesting and memorable, but it can be interesting and memorable in cable by using simpler, more cost-efficient techniques.

FIGURE 7–3. Producing the New Media Message That Motivates

Money-Saving Cable Creative Shortcuts

A major reason for the costs involved in traditional broadcast commercial production is the time needed (or taken) to develop an idea and execute the finished product. For example, a project might require 16 weeks from storyboard development to editing and post-production.

For cable, there is the luxury of neither time nor money. Money-saving shortcuts begin with the selection of a production house that has probably had experience with retail commercials and/or industrial films. They will understand what it means when you want to deliver a package of five cable messages for under $25,000.

In producing advertising for cable, you should:

1. Use relatively simple sets.
2. Use on-location shots, if convenient, but only if a minimum of make-ready is required.
3. Shoot several advertising executions back-to-back.
4. Use smaller casts.
5. Hold editing to a minimum.

And, on the subject of editing, it may be possible to use advertising material you already have in-house by:

1. Adding a tag to existing commercials.
2. Re-tracking the sound track of an old commercial.
3. Re-tracking *and* re-editing two or more commercials into one.
4. Cutting down and restructuring an existing industrial film.

FIGURE 7–4. Traditional Broadcast Production Schedule

8 weeks — storyboard development by agency
2 weeks — legal approvals by agency, advertiser, networks

| Selection of Producer |

2 weeks — casting and location hunt
1 week — filming
3 weeks — editing and post-production

16 weeks — from start to finish

The Quaker Oats Company did this when they sponsored the entire opening night of CBS Cable on October 12, 1981. Their advertising ran thematically through the evening much like a magazine.

In a segment on Mike Nichols's investments in Arabian horses, there was a Ken-L-Ration 83-second spot about a little lost dog. It utilized commercial footage shot for a television special 10 years earlier. Since the product label had been redesigned from when the commercial was first shot, a new tag was simply edited in at the end of the message.

A segment on Liz Swados's dance company, of interest to parents of young children, carried a Fisher-Price billboard plus a 104-second commercial taken from a half-hour Fisher-Price film showing how their toys were designed.

An Alec Guinness drama segment included a 38-second spot on the origin of the Quaker-Man logo, again re-edited from a special commercial produced 10 years earlier.

Another way to hold production costs down is to produce multiple versions of a commercial, some for broadcast and others for cable use. For example, Chevrolet developed a special 90-second avant-garde hard rock music spot with visual collages to run on MTV: Music Television as well as on USA's "Night Flight." But they also ran it on ScreenVision in movie theaters and developed 30-second and 60-second versions for late-night broadcast television. In a single production effort, three versions of an advertising message were developed for cable, theaters, and broadcast television.

Another example of editing and producing for cable involved

"Kawasaki Sets You Free," a cable commercial produced by J. Walter Thompson USA as part of a developmental program to show clients the kinds of commercials that would be compatible with MTV: Music Television. The production included use of existing motorcycle footage, stock outdoor footage, and the development and recording of an original "heavy metal" rock and roll song. This two-minute message was produced as an experimental prototype to take advantage of "Life-Style Association" benefits of Warner-Amex's MTV: Music Television. By using existing footage, production became largely an editing job with the completed project costing only $5,000.

As advertisers investigate new and different ways to produce cable advertising that sells, they should seek out services providing stock sound and video effects that can be integrated into their messages. Among the companies offering such services is Thomas J. Valentino, with a production music library of over 4,000 selections, a sound effects library of over 1,000 effects, and "Video Stock Shots" that include everything from a desert sunset to a forest in winter and from walking in New York to walking in space. Used skillfully, these effects can add to the impact of a cable message. If they are just "thrown-in" for effect, however, they may very well create a negative effect. The key to their use is skill.

Planning Local While Thinking National: The "Split-60"

In another chapter I will discuss cable advertising at the local level and how it may be bought, sold, and created. At this point, however, it must be emphasized that in planning and developing cable advertising at the national level, marketers should consider how these messages might be used locally for tie-ins with dealers, retailers, and franchisees. A "Split-60" commercial is one means by which the national advertiser can accomplish this. It represents two 30-second commercials that run back to back and create the impression of being a single message from a single advertiser. When run on a national satellite network, they appear as one unit, perhaps with two themes or appeals, from the advertiser. Local cable systems are advised of the schedule so that the advertiser's dealers, retailers, or franchisees are able to tie-in a specific localized message. Where this has occurred, the local advertiser's message runs over the second half of the national 30/30. In these cases, the national advertiser's first 30-second commercial stands alone without any creative impediment and leads into the local message.

The possibilities for these kinds of commercials are nearly limit-

"Kawasaki Sets You Free" is an example of the kind of inexpensive long-form commercial that can be developed for a specific program environment.

less. For example, Century 21 might run a "Split-60" on ESPN. In markets where a local Century 21 office wanted to tie-in, it could override the last 30 seconds of the modular unit with its announcement. Or General Foods could feature some of its products in a "Split-60" on Hearst/ABC's "Daytime" and make available the last 30 for overriding by a local grocery chain. A *warning!* This approach will become more common in the future as an increasing number of cable systems purchase the equipment to allow them to insert local commercials. Until then, its application will be limited.

Tips on 800 Numbers in Cable Communications

In developing advertising for cable, many companies have an interest in building a direct response mechanism into their messages. The Direct Mail Marketing Association can provide considerable insight as to how cable and 800 numbers can work together most effectively. Here are five of their suggestions:

1. Go to a good direct response advertising agency. It will know how to buy the time properly and to create a commercial that *sells*.

2. Properly utilize the 800 number in your ad. It should be displayed for at least 15 to 20 seconds.

3. Make sure there are adequate facilities to handle the 800 number, whether you are using an in-house facility or an outside telephone marketing service bureau.

4. Make sure that your 800 number has "rhythm" so that the number is easy to remember. For example, 800-228-8000 is far better than 800-647-2594!

5. Use a variety of 800 numbers for source coding. This will enable you to trace where your calls are coming from and to determine which cable effort is working best for you. This is especially important today with the very limited amount of available audience research data.

Understanding and Creating the Infomercial

What is the infomercial? A broadcast commercial seeks to generate awareness for a product or service, to build brand identity, and to register usually a single major sales point. In contrast, an infomercial, through the luxury of time, can explain the product's or service's benefits, uses, and characteristics. It can also provide interested viewers with the means of securing more information, a sample, or the product or service itself. As Bill Harvey, publisher of the *Media Science Newsletter*, pointed out, "By involving the viewer more in the message, an infomercial can be regarded by the viewers as more relevant, absorbing, sincere, and real than a short-form broadcast commercial."

All products and services were not meant to be sold in the confines of 30- or 60-second television messages. Unfortunately, the structure and pricing of broadcast television has generally necessitated this. Infomercials, however, allow the marketer to break out of this mold and provide the more detailed selling information he would like to give in broadcast television but cannot afford. And, because of the selective audience appeals of many cable networks or local system channels, he can also deliver the message to a more targeted rather than to a mass television audience. In a nutshell, infomercials can provide advertisers with the in-depth communications ability of print, coupled with the video impact of television.

Producing Effective Infomercials Efficiently

While a 30-second broadcast commercial may cost well over $100,000 to produce, an infomercial can be delivered for well under $5,000. Its aim is to impart information, and this helpful service aspect, rather than expensive production techniques, should hold the viewer's attention. Costs are held down by shooting several infomercials on a single day (rather than one over several days), using stock footage where available and applicable, editing on the spot, using tape and not film, and employing a minimum number of cameras *and* a minimum number of creative and production people.

At the local level, an infomercial would be even simpler to produce if modular material supplied by the national advertiser was used and/or if commercials were shot on location at the local retailer, dealer, or franchisee's place of business. In this case, it might, for example, cost only $100 to go into the local lawn and garden store and shoot an infomercial on how to select plants for your yard.

Unfortunately, an infomercial can have fast viewer wearout and limited audience reach. Therefore, an advertiser might find it necessary to produce more of them than he would 30-second commercials for a broadcast pool. This is another reason why the cost of producing each infomercial must be held down.

Guidelines for Creating Infomercials That Sell

Research has generally indicated that infomercials should be well received by viewers. They are very receptive to the idea of helpful, informative, long-length messages and often point out that they would

FIGURE 7–5. The Infomercial

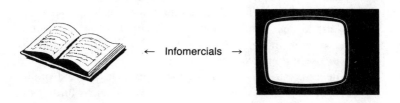

← Infomercials →

like the kind of buying information that "old-time salesmen" used to provide. Unfortunately, salespeople today generally aren't what they used to be, and perhaps that is why there may be this appetite for video infomercials. One warning: don't think that all you have to do to produce an infomercial is stretch a 30-second commercial to two minutes or so. Viewers will recognize this immediately as nothing more than a long-form commercial and will be turned off.

Here are some guidelines for the creation of effective, selling infomercials:

1. Remember that the job of the infomercial is to provide depth of impression, not merely broad exposure of a message. It must provide the viewer with genuinely helpful information.

2. The infomercial gives you the time to tell the viewer what you are going to tell him about, to tell it to him, and then to tell him what you have told him!

3. Be sure to explain to the viewer how to go about getting more information on the product or service. *And* make it easy for him to do so.

4. Provide a built-in direct response mechanism where appropriate.

5. Avoid exaggeration and puffery.

6. For added effectiveness, consider using short-form, tie-in commercials that can run on cable networks or individual cable systems and reinforce specific points in your long-form informational messages. For example, five- to seven-minute infomercials by a tax service like H & R Block could provide in-depth information on how to maintain tax records and assemble your material each year at income tax time. A series of 30-second "mini-messages" could focus on separate tips on how to maximize your savings in specific areas.

7. Unlike broadcast commercials, which may be exposed up to 10 times without any deleterious side-effects, infomercials tend to wear out fast. A single exposure may be sufficient to convey its intended message. Hence, if an advertiser intends to use infomercials over an extended period of time, it is necessary to produce a sufficient number to avoid undue wearout of any single message. The problem is further exacerbated by the relatively low reach of an infomercial against

its selected cable audience relative to the broader reach of a mass-audience broadcast schedule.

8. An infomercial should have a spokesperson. He need not be an actor and preferably should be someone close to the product at the client company. If the product manager, R&D manager, or even the president of the company can present well, he or she will make a very creditable spokesperson for the message.

9. If you have produced an infomercial, extend its life by making it available for showings at sales meetings and distribute it on tape to dealer organizations.

An excellent example of the development of an effective infomercial involved Sealy, the bedding manufacturer, and the Modern Satellite Network's "Home Shopping Show." A half-hour infomercial was produced in three segments that could be used together or shown individually. This consumer information program included Sealy's director of medical research, the director of the Center for Pain Studies at the Rehabilitation Institute of Chicago, the director of product development, and the director of marketing. Sealy had "freedom of time" to discuss the importance of selecting quality mattresses to ensure proper rest, good health, and a "sense of wellness." Obviously, none of this could have been done in only 30 seconds.

FIGURE 7–6. Infomercials: Long and Short

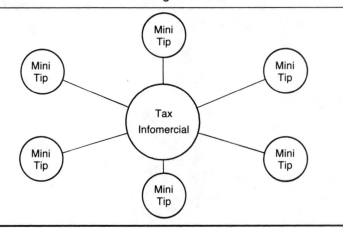

Tapes, Discs, and the Future

As we move from the early to the mid-eighties, attention will be focused not only on cable, but on the development of appropriate forms of advertising on video tape or video disc. Service/product areas such as baby and child care, gardening, exercise and fitness, travel, home computers, and photography are just a few examples of areas in which there is a high degree of either general or specialized consumer interest and where advertiser-related tapes or discs might be meaningful commercial vehicles. Bill Harvey offers five excellent tips on producing tapes and discs that should be considered by anyone investigating this media approach:

1. It should not be something you can get on television.
2. It should require or encourage re-watching.
3. It should look good on wall screen as well as on conventional televisions.
4. It should appeal to movie and record buffs as well as videophiles.
5. It should be produceable without an enormous movie budget.

The point is that *all* of the new media can and probably will be advertising supported to at least some degree. It will require imagination and creativity to develop advertising that will most effectively communicate a marketer's messages, regardless of whether they are carried via cable, cassette, disc, videotex, or any distribution system not even yet thought of. This will be the advertising creative challenge of the years ahead.

CHAPTER 8

Measuring Results: Qualitative and Quantitative Approaches

The Need for *Facts* . . . Not Just *Faith*

The dramatic growth of the new electronic media has been accompanied by an increasing interest in cable advertising. But relatively little has been available in terms of hard, actionable research data. Moreover, most advertisers have been willing to spend their dollars in cable largely on a matter of faith that they would receive in viewer response a value approximately equal to the money they have invested.

This is in sharp contrast to the buyers of magazines, newspapers, outdoor, radio, and television. They evaluate their media purchases on the basis of data from MRI, SMRB, ABC, Arbitron, Nielsen, TAB, and many other research and auditing firms. Today, however, as cable advertisers consider larger investments in the medium and as many of them investigate extending their involvement from a national to a local basis, they must justify these expenditures of their money. It is no longer just experimentation. They need confirmation of the viewing audiences.

What Is Available Now

As of today, cable audience information comes from a variety of sources, but there are really no standard measures to cover the entire medium on either a national or local level.

On a national basis:

- Qualitative profile information on cable's general characteristics is available from Arbitron's New Electronic Media Study, MRI, SMRB, VideoProbeIndex, and other omnibus services.
- Several times a year, the Nielsen Television Index provides national meter data on cable and pay cable usage overall.

TABLE 8-1. Who Views Cable, Pay Cable, and STV
Adults 18-Plus, Indexed against General Population

	General Population	Cable	Pay TV	STV
Age				
18–24	100	88	101	173
25–34	100	109	120	91
35–44	100	111	138	90
45–54	100	116	117	102
55+	100	86	54	68
Education				
Graduated college	100	115	127	80
Attended college	100	112	118	131
Graduated H.S.	100	101	102	88
Did not graduate H.S.	100	84	73	108
Household income (000)				
$25+	100	128	153	113
$20–$25	100	112	112	105
$15–$20	100	95	88	77
$10–$15	100	81	68	97
$5–$10	100	76	61	90
Under $5	100	56	26	105
Census region				
Northeast	100	118	118	88
North Central	100	84	86	84
South	100	101	95	82
West	100	100	106	168
County Size				
A	100	75	98	133
B	100	110	109	103
C	100	132	107	52
D	100	107	81	69

- In terms of specific advertiser-supported satellite networks, WTBS was the first to be regularly measured by Nielsen's national meter service and was joined in the summer of 1982 by the Cable News Network.

- The Nielsen Home Video Index provides broad, diary-based, descriptive data of trends in cable and pay cable viewing levels. (As is characteristic of the diary, these cable audience levels are substantially lower than those shown by Nielsen's meter.)

TABLE 8–1. Who Views Cable, Pay Cable, and STV (continued)
Adults 18-Plus, Indexed against General Population

	General Population	Cable	Pay TV	STV
SMSA				
Central City	100	101	107	106
Suburban	100	89	108	120
Non-SMSA	100	111	81	68
*Purchases in past 12 months**				
Cable	100	347	462	320
Pay TV	100	347	651	401
Product usage				
Beer	100	99	113	122
Deodorants	100	101	102	107
Chewing gum	100	105	107	111
Cold remedies	100	109	117	124
Packaged candy	100	107	100	143
Carbonated beverages	100	99	106	112
TV viewers				
Total week				
Heavy viewers	100	106	104	139
Light viewers	100	90	86	70
Daytime				
Heavy viewers	100	100	97	139
Light viewers	100	104	104	85
TV prime				
Heavy viewers	100	109	114	124
Light viewers	100	90	83	86

*Includes second purchases.

Source: MRI, Fall 1981

- Arbitron CAMP surveys and the Nielsen Home Video Index can provide special studies of satellite network audience delivery in specific local areas based upon telephone coincidental and telephone recall methods.
- Arbitron's National Network Cable Report measures the audience to eight satellite cable networks based upon telephone recall.

At the local market level, diary and telephone recall studies can provide information on a special order basis. Companies providing this service include Arbitron, Nielsen, VideoProbeIndex, Communications Marketing, Information & Analysis, Media Science Measurement, and others.

What Is Needed

The significant point is that audience research data are available today on over 150 consumer magazines, newspapers in some 50 markets, and all broadcast radio and television stations. The amount of information available on cable, however, has been limited to very few studies issued *only occasionally* on *different cable services* using *different bases* and *different measurement techniques.*

What national advertisers and agencies are seeking is comparable data on the larger advertising-supported satellite cable networks utilizing the same data bases and coming out several times a year. And at the local level, cable operators and marketers interested in the placement of local cable advertising need methods of securing some form of audience research on more than just a sporadic basis. The big question is, How does the industry get from where it is to where it should be?

A Question of Technique: What to Measure

Back when the earth was flat and people didn't know what lay beyond the horizon, everyone was afraid to take chances and venture into the unknown. Then one day we learned the earth was really round. Everyone wanted a piece of the action, and Hertz Rent-A-Boat could barely keep up with the demand. Explorers went out looking for something, and many were not even sure what they were seeking.

It's now 500 years later, and we're still in the era of exploration.

This time, however, instead of looking out over an ocean, we're treading water in a sea of cable confusion. On one hand, we're trying to come up with ways of measuring the audience of cable while, on the other hand, only a small number of explorers really understand what the medium is all about.

The initial problem is really not how to measure cable. It's *what* to measure. The problem is to get our explorers' crews to understand that cable is not just a lot of additional low-rated television channels, but an entirely new medium. (In other words, the cable world is not flat. It's round. And, in fact, as scientists discovered about the earth, it may even be oval!)

Cable gives us an opportunity to decide exactly what it is out there we want to measure before we start counting heads.

Learning from the Past

Too often in the past, research has dictated how we should evaluate media rather than the reverse. Just ask any media planner:

1. Why do we talk about four-week television reach?

2. Why do we examine radio on a one-week cume basis?

3. Why do we compare magazines on a total audience basis?

Many either won't know or will think that it's based on knowledge of how marketing and media influence each other. In reality, the answers are that:

1. Nielsen initially had six analysis periods each year. They picked the four center weeks of each two-month period to report on. That's how we got four-week reach.

2. We examine radio on a one-week cume basis because diaries are kept for a week.

3. Total issue audience became gospel largely through the efforts of *Life* magazine to come up with big numbers.

The problems in magazine measurement should be a lesson as we look ahead to measuring cable. We have modified through-the-book, recent reading, modified recent reading, calibration, and cover recognition, . . . and only now are we really acknowledging that all of them may be right and that they just may be measuring different things.

In the past, we have forced media into existing measurement techniques, expecting the measurement devices to be something they are not. We have a clean slate with cable, and the first task must be to understand just what it is and what it is capable of doing before we try to measure it.

Reflecting How Cable Is Used

From the outset it is apparent that any cable audience research methodology must reflect how people view cable and how advertisers should buy the medium. For example, cable networks or local systems might sell by dayparts or on the basis of individual programs or combinations of programs. In some instances, audience data that report daily or weekly cumulative circulation levels across dayparts might be sufficient. In other cases, where cable networks or large interconnects reached very large numbers of homes and where commercial time is purchased much as it is for broadcast television, individual rating information would be important. And in still other instances, there might be more interest in general qualitative data on the characteristics of the audience, similar to that available for newspapers and magazines.

A First Step—Counting Cable Households

The Problem of a True Count

One of today's more perplexing problems in cable measurement is simply getting an accurate count of the number of cable households in existence. At any given point in time, in any given area, there may be anywhere from a 10 percent to 20 percent difference in the number of cable households, depending on whether the data come from subscriber counts or actual house-to-house censuses. The differences are due to:

- Piracy. Households are buying illegal convertors and hooking up to the cable themselves.
- Homes that have cancelled their subscriptions, but have not yet been disconnected.
- Large apartment and condo complexes where the building management does not report the full number of individual unit subscribers.
- Homes that have subscribed but have not yet been included in the subscriber count.

Some experts estimate that as many as 1 out of every 10 cable homes is currently receiving its signal illegally by hijacking signals with homemade antennas, by buying illegal convertors and bypassing in-home decoder boxes, or by climbing telephone poles to knock out security traps.

Obviously, a problem in cable audience measurement begins simply with knowing how many cable homes are out there.

From Cable Count to Coverage Area

Before one attempts to measure the viewing audience of specific cable networks or local cable systems, it is necessary to establish a more basic audience measurement that is just one step removed from the cable household count. This is the "coverage area" of the potential cable advertising vehicle. It represents these cable homes located in areas that can receive the cable service. Advertisers will want to examine the distribution of homes served by a cable network or cable system in relation to their own sales territories or marketing areas. For example, the Cable News Network provided the information shown in Tables 8–2 and 8–3 on the distribution of their March 1982 household count by Nielsen territories. From these data, an advertiser can see where the service's audience is concentrated or if it is distributed in line with population or other measures.

Advertisers should be very careful when it comes to examining cable household distribution or audience data within customary television universes such as ADIs or DMAs. Arbitron's Area of Dominant Influence (ADI) and Nielsen's Designated Market Area (DMA) were developed in the 1960s to define areas with homogeneous television station availability. The entire country was divided into approximately 200 ADIs or DMAs, and each area reflected all of the counties that devoted a majority of its viewing time to stations in the home county. The New York ADI, for example, encompasses 30 counties.

Bill Harvey has developed a concept of the Area of Community Influence (ACI). It follows the lines of the cable franchise area so that a given ADI might contain as many as 100 different cable systems.

TABLE 8–2. CNN Distribution by Nielsen Territory

Nielsen Marketing Research Service Territory	CNN (Percent)	Total U.S. TV Households (Percent)
New England	5	5
Mid Atlantic	18	18
East Central	10	15
West Central	14	17
Southeast	19	17
Southwest	14	11
Pacific	20	17
Total	100	100

Source: Cable News Network, March 1982.

TABLE 8–3. CNN Compared to All Television Households

	CNN (Percent)		Percent Total U.S. TV Households	
"A" Counties	34	> 71	38	> 66
"B" Counties	37		28	
"C" Counties	20		18	
"D" Counties	9		15	

Source: Cable News Network, March 1982.

The ACI concept not only allows an advertiser to determine how many cable systems must be bought to cover a given marketing area, it also more accurately pinpoints the homes that would be reached by buying cable on a specific number of individual cable systems.

Some Basic Problems Facing Cable Measurement

Over the years, a number of methods have been developed to measure television viewing audiences. The most common methodologies have involved diaries, telephone interviews, and meters. In broadcast television, the cost of these services generally has been shared among several stations in a given local area. Because cable operators have exclusive franchise areas, however, the expense of audience research for a single local system must be borne by that system itself and cannot be shared. In addition, cable, both on a local and on a national basis, represents a medium of many more channels than in the case of television and with far lower ratings. Furthermore, audiences of most of the national satellite networks are distributed in a non-uniform pattern across the country, often making it extremely costly to develop a sample of viewers. And, of course, at the local level, many system owners are simply totally unaccustomed to research, have not involved themselves in the sales of local advertising, and must be convinced that it is to their benefit to provide audience data.

Taken together, all of these factors point to the need for the development of methods of reliable, low-cost audience measurement that can be effectively and efficiently used by both advertisers and the cable industry in buying and evaluating cable advertising concepts.

Five Basic Audience Measurement Methods

Five methods presently exist for measuring cable audiences—diaries, meters, telephone coincidentals, telephone recall, and two-way interactive cable.

In the long run, two-way interactive cable will provide instantaneous feedback as to exactly what cable channels are tuned and in what homes. It may be many years, however, before a sufficient number of them are distributed throughout the marketplace to provide the needed sample frames. And, of course, even with a two-way interactive response device that does not rely upon reporting by an individual, there is the problem that it measures set tuning rather than individual viewing.

While meters may well emerge as the most effective means of measuring the audience of national satellite networks, they, like two-way interactive cable, are not 100 percent adequate. They do not now reflect who is viewing the cable service, although tests are underway to develop a "people" as opposed to a "household" meter. And, from a local standpoint, individual meters are extremely costly. Today, they are used for local broadcast measurement only in a limited number of markets, and it is doubtful that they will ever expand to the measurement of individual market cable data.

Telephone recall techniques are used to gather data on viewing anywhere from today to yesterday or to the previous seven days. In general, the longer the recall period, the greater the influence of memory on the accuracy of the response. Hence, it is generally regarded that recall is of greatest value in collecting data on audience levels over a period of time rather than on individual quarter hours.

Since memory is not involved in telephone coincidental surveys, they generally have been regarded as the most accurate method of gathering data on who is viewing now. From a cost standpoint, however, very large samples are needed, since every time period measured requires a sample sufficient from which to project results.

From an economic standpoint, diaries are a very efficient method of measurement. However, they rely on the diarist's memory, and the element of memory has typically been responsible for showing cable audiences viewing levels that fall well below those reported either by meters or by coincidental methods. In broadcasting, it has long been acknowledged that diaries tend to underreport the audiences of lower-rated fringe and daytime programs as well as audiences of independent TV stations, ethnic stations, and public broadcasting stations.

With these problems, some might ask, why has the diary been used? The answer is simply that it has been the most economical method to gather viewing data for an entire week from an individual household. One of the reasons for the research proposed by the Ad Hoc Cable Audience Measurement Committee (discussed later) was to determine whether the diary could be satisfactorily implemented to measure the audience of cable.

Any decision as to which cable research method to use must take into consideration the cost of the method used to get the data generated. As might be expected, the more accurate the data generated, the more expensive the technique. (See Table 8-4.)

The Need for Cable's *Own* Measurement Systems

Because cable audiences are so fractionated and the numbers far smaller than for broadcast television, it is generally acknowledged that special studies are required to measure them. They cannot be adequately picked up in existing Arbitron or Nielsen television rating reports. While both Arbitron and Nielsen have developed ways of reporting broad market audience totals of cable audiences in their

TABLE 8–4. A Cable Research Comparograph

Methodology	Description	Data Yielded	Approximate Cost (1982)
Interactive Cable	Requires diaries placed in field to determine demographics behind specific viewing.	Tracks exact TV set usage through census or sample.	Cost of scanning device, computer software and, if required, an impartial third- party audit.
Meters	Requires diaries placed in field to determine demographics behind specific viewing.	Tracks exact TV set usage of sample.	In excess of $300,000 a year for 500 meters.
Telephone Recall (24-hour)	Asks additional questions for programming research. Relies on participant's ability to recall viewing.	Daily cume for each channel by daypart, and weekly cume for each channel. Demographics broken down into sex/age categories.	Starts at $3,000 for 200 completed phone interviews over a seven-day period.
Telephone Coincidentals	Simple formula of questioning household on viewing at exact moment of phone call. Accurate but limited information.	Average quarter-hour ratings and distribution of audience in terms of men, women, teens and children.	$300-$450 for over 200 completed phone calls to measure particular program or channel.
Diary	Follows household viewing across a week's time. Assertion is that method understates programming of limited viewing.	Average quarter-hour ratings, weekly cume data, and numerous demographic breakdowns.	$8,000 for 250 diaries (price will differ if tabulations are made from existing diary base for broadcast research).

regular reports, these generally are insufficient for evaluating the medium, and, in most cases, very little viewing data show up. This is because the cable audiences are related to an entire television coverage area rather than to only their own subscriber base.

If, for example, a cable system covers 20 percent of a total market, audience data for the system must be measured and shown as a percentage of its own individual franchise area, not as a percentage of the entire television marketplace.

The general consensus is that cable audience data also cannot be measured in the same diaries that presently are used to gather broadcast audience information because:

- By and large there is not adequate space for listing all of cable's multiple channels.
- There is a general confusion on the part of the viewer as to identification of the different cable channels.
- Just as it has long been felt that diaries tend to penalize the lower-rated, occasionally viewed broadcast programs, there would be an even greater problem with cable's small audience levels.

To a certain extent, the problems in measuring cable audiences may be similar to those involved in radio broadcasting measurement. Radio broadcasters have long been concerned about how to measure fractionalized listening to 30, 40, or more stations in a market. With cable, these numbers of channels will be commonplace. For example,

FIGURE 8–1. Correct Base for Cable Audience Measurement

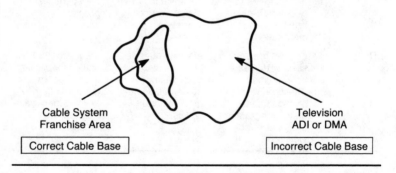

Cable System Franchise Area

Television ADI or DMA

| Correct Cable Base | Incorrect Cable Base |

Table 8-5 shows Viacom's Greendale, Wisconsin, system as of March 1982 when 60 out of a possible 82 channels were being programmed.

The Ad Hoc Committee

The Ad Hoc Cable Measurement Committee was formed to coordinate the development of research methodologies best suited to the unique characteristics of cable. The Committee's objective was not to develop a rating system per se, but to spearhead the evaluation of various methodologies for measuring local cable audiences, and to determine which methodologies might represent the most viable, reliable, and cost-effective method or methods for obtaining viewing data for programming and advertising purposes. The Committee was made

TABLE 8–5. Multiple Channels to Measure

Channel	Programming	Channel	Programming
2A	Chicago CBS	3B	Library Access
3A	Promos	4B	Educational Access
4A	Milwaukee NBC	5B	Local Government
5A	Chicago NBC	6B	Consumer News
6A	Milwaukee CBS	7B	Local Origination
7A	Chicago ABC	8B	Government Access
8A	News Features	9B	Public Access
9A	Chicago (WGN)	10B	Time/Weather
10A	Milwaukee Educational	11B	Bulletin Board
12A	Milwaukee ABC	12B	Swap & Shop
17A	Atlanta (WTBS)	13B	Government Access
18A	Milwaukee Independent	17B	Showtime
20A	ESPN	18B	HBO
21A	WHA (Educational)	20B	The Movie Channel
22A	Job Bank	21B	Cinemax
23A	C-SPAN	22B	HTN
24A	Chicago Independent	23B	Bravo
27A	Financial News	24B	Galavision
28A	New York (WOR)	27B	Swap & Shop
30A	CBS Cable	28B	Nickelodeon
32A	Chicago Independent	30B	CBN
33A	National Jewish TV	32B	USA
36A	Milwaukee Educational	33B	CNN
37A	MSN	34B	National/Internat'l News
38A	PTL	35B	Sportswire
39A	NCN	36B	State/Local News
40A	Trinity	37B	AETN
41A	BET	39B	SIN
2B	Channel Guide	41B	MTV

up of representatives from major firms in both the cable and advertising fields and gathered as many views as possible regarding local cable advertising and sales methods. This effort was the first time that the buyers and sellers of advertising time cooperated in sponsoring such a project.

In early 1981, the Ad Hoc Committee issued its report. It recommended testing several alternative measurement methods for local cable audience measurement. The A.C. Nielsen Company was selected to conduct this Cable Audience Methodology Study (CAMS), which was conducted in June 1982 under the supervision of the Industry Research Standards Committee, jointly sponsored by the NCTA and the Cable Television Advertising Bureau.

Seven separate tests were conducted to gather information on household viewing and to determine personal cable usage. Diary and telephone collection techniques were evaluated, and both daypart and quarter-hour viewing levels were tested. All of these data were validated against a series of telephone coincidental surveys, which, for many years, have been accepted as the most reliable way to establish accurate audience levels. As this book went to press, the results of this test were just being tabulated.

The Creative Cable Research Kit

The needs for research vary greatly depending on who is using it and what they are using it for. For an advertiser making a direct-response offer, research is built in. It's the response delivered.

For most advertisers, however, this is not the case. An advertiser who spreads advertising messages throughout a cable channel on various days and at various hours may well find that a cumulative audience measure of those people who view these time periods at one time or another may be adequate. If, however, that advertiser selects specific programs at different times—where there is not a continuity of viewing—there would be a greater need for individual program rating data. For some advertisers (for instance, local advertisers who are presently using local shopping newspapers and have that kind of information now available to them), it may suffice to have an adequate measure of the distribution of the cable homes and the characteristics of the subscribers.

They say that necessity is the mother of invention, and, in lieu of regularly reported audience data, many advertisers, agencies, cable networks, and system operators have been extremely imaginative. An advertiser can even put together a "Creative Cable Research Kit" by investigating the variety of efforts thus far. Some examples follow.

The "Daytime" Woman

Before the Hearst/ABC "Daytime" service went on the air in March 1982, their research director sought to identify its potential audience. The objective was to define who the "Daytime" woman was and to take a look at her demographics, psychographics, activities, and consumer habits.

Hearst/ABC examined "Daytime's" Projected Viewing Areas— those municipalities that were carrying the SATCOM I Transponder 22 signal from 1:00 to 5:00 P.M. EST (the time period to be occupied by "Daytime"). The PVA's were identified by five-digit zip codes that were matched against the zip codes on the Simmons Market Research Bureau's 1980 data file.

The characteristics of a representative sample of women who were cable subscribers living in the PVA's were then compared to a sample of all adult women nationally. Hearst/ABC used this comparison to show how the "Daytime" woman compared favorably in demographics, psychographics, and consumer behavior with all women. (See Table 8–6.)

Research like this must be examined very carefully. It assumes

TABLE 8–6. Characteristics of "Daytime" Female Cable Subscribers Compared with All Adult Women

	"Daytime" Percent	National Percent	"Daytime" Index
"Daytime" Household:			
Household Size 3+	64.0	54.0	118
Number of 18+ adults	30.0	26.0	115
Presence of children	51.0	44.0	116
Married	66.0	61.0	108
Married 10 years+	75.0	70.0	107
Live in top 25 ADI's	61.5	53.8	114
"Daytime" Woman:			
25–49	47.6	44.8	106
Attended college	32.3	27.7	116

Source: Simmons Market Research Bureau

that the "Daytime" viewing audience will actually parallel that of women living in areas that receive the "Daytime" signal. Such may not in fact be the case. Nevertheless, it is an imaginative effort to provide projected audience characteristic data when none are available, and it can be a very effective sales tool from the standpoint of the cable network and cable system.

CBS Cable and the Upscale Viewer

In a project similar to the "Daytime" study, CBS Cable wanted to show the upscale nature of their audience in November 1981, at a time when audience information on it was not yet available. As with "Daytime," CBS Cable used Simmons Market Research Bureau data. In this case, they analyzed the demographics, product purchase habits, and media usage of households that reported viewing cultural and performing arts television programming in the Simmons National Study of Media and Markets. CBS Cable stipulated to Simmons that all households used in this analysis had to be located within the coverage area of cable systems carrying their new cultural network. The results showed that these viewers were affluent, professionally successful, and well-educated and that they were consumers of expensive products and services and regular readers of upscale print media. For example, Table 8–7 shows their reading preferences compared to the national average.

Once again, the assumption is made that the CBS Cable viewing audience will have the same characteristics as those people in

TABLE 8–7. Reading Preferences of Audience in CBS Cable Area

Publication	Percent Greater than U.S. Average
Business Week	+38
Forbes	+32
Fortune	+87
Scientific American	+45
U. S. News	+49
Wall Street Journal	+43
Any Epicurean Publication	+49
Omni	+84
New Yorker	+83

Source: SMRB Special Analysis (November 1981)

Simmons national diary sample who are located in the coverage area of cable systems carrying CBS Cable and who view cultural and performing arts programming. In this case, the assumption is probably fairly valid, and it does provide a cable network and potential advertiser with some relevant research data where none would otherwise exist.

Who Watches the "Home Shopping Show"?

In some cases, research can be generated by a follow-up questionnaire to people who respond to direct response offers. The Modern Satellite Network's "Home Shopping Show" did this in a survey conducted by the A.C. Nielsen Company. A questionnaire was mailed to 1,000 people who had used a toll-free telephone number to respond to offers made by advertisers on the "Home Shopping Show."

From 355 returned and completed questionnaires, the following results were tabulated:

- Customers had higher than average household incomes—40 percent over $30,000.
- Respondents were frequent purchasers of appliances, with 26 percent having purchased a television set in the previous 12 months.
- More than half the respondents were women, with 28 percent in the 18-34 age group and 25 percent in the 35-54 age group; 26 percent of respondents were men 18-54.
- Respondents were heavy users of coupons, with 91 percent coming from households where newspaper or store coupons had been used to purchase a product in the previous 30 days.
- Three-fourths came from households in which a member had ordered a product from a catalog in the previous year; 49 percent made more than three such orders a year.
- When asked their reasons for watching the show, 95 percent cited the show's informational value; 52 percent said they watched the show twice a month or more.
- Nearly a third (32 percent) of viewers said "The Home Shopping Show" influenced them to purchase a particular product.

Since the "Home Shopping Show" is telecast during the daytime, some advertisers questioned whether it reached an affluent audience. The purpose of this research by the Modern Satellite Network was to show that the "Home Shopping Show" had a very active, involved audience—that it was not simply a passive daytime television viewing audience. Again, a research technique was devised to produce qualitative data where no regular survey information existed.

Developing a Local System's Profile

At the local level, *Cable Plus*, a weekly cable and network TV listing publication in the Dallas/Fort Worth area conducted a study defining the magazine's audience for advertisers. Of 500 questionnaires delivered to subscribers, 252 were returned. *Cable Plus* used the survey results not only to reflect the characteristics of their reading audience, but also to present a profile of the market's cable audience. They assumed that this was a logical assumption to make since its readers all subscribed to cable. The results showed an affluent audience that devoted approximately 23 percent of its viewing time to the cable networks (46 percent to the premium cable channels) and owned a significant number of video cassette recorders and video games.

The *Cable Plus* study is one approach to gathering information on a local market basis. Another way is for an individual cable system to include with its monthly bill a questionnaire on viewer interests and behavior.

Polling via Telephone

No one would question the fact that if television programmers knew, before putting a show on the air, whether people would want to watch it or not, they would stand a better chance of getting a high rating. WTBS and Bristol-Myers did just this when they allowed viewers the opportunity to select the February 1982 "Bristol-Myers Theater" motion picture via AT&T's "Dial-It Telephone Polling System."

During the January 4 Bristol-Myers movie, viewers nationwide were told that they could call a designated telephone number to vote their preference on a given film for February. There were six possible movies corresponding to six different numbers viewers could dial to indicate their preferences. Results were tabulated instantly by AT&T and announced by WTBS personality, Bill Tush, immediately upon the conclusion of the January 4 film. "Paint Your Wagon," the movie that received the most votes, was then scheduled for February 1. (Overall results are shown in Table 8–8.)

Recipes, Cookbooks, and Research

Analysis of the distribution of response to the offers can also provide

TABLE 8–8. Results of "Dial-It" Telephone Poll

Movie	Response
Paint Your Wagon	1,329
Damn Yankees	804
Flower Drum Song	747
Angel in My Pocket	627
Greatest Show on Earth	577
Night and Day	480
	4,654

some guidance as to the distribution of the viewing audience. This is actually a very old technique. Back in the early days of radio, before the development of audience measurement services, Procter & Gamble offered flower seeds on one of their radio shows. By analyzing where the responses came from, they had some idea of where the radio show's listeners were.

Kraft used a response mechanism to determine if—in lieu of traditional rating data—there was anyone out there watching. They didn't believe the people who said, "If you can measure something in a rating report, it exists, and if you can't, it doesn't exist." Kraft offered a free recipe book to anyone who wrote to WTBS and the Cable News Network. The cable advertising schedules were small and ran only three weeks. Kraft had no solid rating data when the buy was made and didn't know whether the offer would pull 100 or 1,000 requests.

The main point of the test was to see if, lacking this solid audience data, there was anyone out there to jiggle the meter. In total, over 10,000 requests came in. Considering that the offer was a modest one and viewers had to be alert to catch it, this response was outstanding. Kraft found that an active, alert audience did exist for cable even though they couldn't measure it in traditional terms. And to gather even further data, they surveyed a sample of the respondents with a simple postcard questionnaire asking about their product usage. (See Figure 8-2.)

A Homegrown Survey of a Single System

According to a Leo Burnett Company study of advertising-supported cable networks, MTV: Music Television had attracted a substantial audience only six weeks after its debut in August 1981. A telephone survey by Leo Burnett's Media Research staff of cable subscribers to a single system in McHenry, Illinois, showed that familiarity and viewing levels for MTV: Music Television equaled or surpassed that of older services such as CNN, ESPN, and the USA Network. While this was a relatively small survey in only one county, later research by MTV: Music Television substantiated the results.

A Cable Subscriber Study

A cable system operator knows that to maximize profits, it is necessary to acheive the highest possible subscriber base with the lowest possible monthly "churn" (or turnover). To do so, cable operators must be ever mindful of the needs of the community they serve and offer services that the community wants. This is especially true today, where the majority of all systems are still only 12 channels, and a relatively small number have more than 20. It is important that these systems constantly assess the interests of their communities in order to know which cable services to add and which to drop.

Gabe Samuels, Vice President of Media Resources and Research —J. Walter Thompson USA, developed a format for a simple sub-

---

FIGURE 8–2. Kraft Cable Questionnaire

Would you answer a few questions for us?
and please drop this card in the mail.

1 Often food companies such as Kraft feature recipes which specify their brand(s). The first time you try one of these recipes how likely are you to use the specified brand?
( ) **Almost always**   ( ) **Usually**   ( ) **Sometimes**   ( ) **Never**

2 Now, when you use the recipe a second or third time, are you more likely to continue to use the specified brand or are you more likely to make a substitution?
( ) **Use specified brand**   ( ) **Make a substitution**

3 Please look through the recipes in the Kraft booklet you just received. On the lines below, list any Kraft products you regularly buy and that you would use in order to make one or more of these recipes.

KRAFT PRODUCTS:_____

_____

4 On the lines below, please list any Kraft products that you do *not* usually buy, but *would* buy in order to make one or more of these recipes.

KRAFT PRODUCTS:_____

_____

Thank you so much

scriber study that local systems can use periodically to assess viewer preferences and service satisfaction. This survey can be adapted to almost any system with questionnaires distributed in the monthly statements or as special mailings. (See Figure 8-3.)

The Old Spice Quiz

In an earlier chapter, I mentioned how Old Spice created a sports trivia contest that centered on events and stars in baseball, basketball, boxing, football, and hockey. Commercials ran in a variety of ESPN sports with each program's responses identified by a different code. Old Spice was in effect developing its own cable rating system before rating data were generally available on ESPN.

The "Advertising Impact Scoreboard"

If a cable system operator is interested in actively soliciting local advertising in a community, he or she will find that a common response from the merchants is, "But how will I know if anyone sees my advertising?" As a promotion idea, the cable system might provide its advertisers with an "Advertising Impact Scorecard" like the one shown in Figure 8-4.

When a customer calls or stops into the store, the owner asks, "How did you find out about our sale (or our special promotion)?" Or, "Where have you heard about us lately?" He checks off each response on the "Advertising Impact Scorecard" and can see at a glance where the advertising response is coming from. *If this seems risky, the cable system operator must remember that, in the long run, cable will succeed as a local advertising medium only if it produces results.* It is up to each cable system to work hand in hand with local advertisers to assure good results by helping them in the execution, promotion, and evaluation of their advertising.

Cumulative Audience: A Helpful Tool

Cumulative audience data can provide a clue as to the broad potential of a cable service to generate audience based on viewing over an extended period of time. The Arbitron National Network Cable Report provided such data in their first viewership study of eight pay and basic satellite programming services. Table 8-9 shows the weekly cume ratings from this November 1981 study.

FIGURE 8–3. Cable System Subscriber Questionnaire

(Date)

(NAME OF TOWN) CABLE TELEVISION
ADVISORY COMMITTEE

Dear (Name of Town) Resident:

Your cable TV committee has been appointed to advise the (City Council, Town Board, etc.) as to the performance of (Cable System). The attached questionnaire is our official request for information from residents of this town.

Since it would have been impractical for us to question every single household in the town or tabulate the answers—we have randomly selected a representative sample of the total population.

We need a 100% response to insure that the results of this survey are representative. It is, therefore, necessary that everybody reply.

For everybody's convenience, we enclose a self-addressed, post-paid envelope in which you can mail your questionnaires back to us.

If you are unable or unwilling for any reason to respond to this questionnaire, please call us at (Phone Number). (Leave a message for the Cable Committee.)

Thank you for your cooperation.

PLEASE RETURN WITHIN 7 DAYS

QUESTIONNAIRE

1. Are you currently a cable TV subscriber?   Yes ☐ No ☐
   If you answered yes, please skip to question 5.

2. Have you previously subscribed to the cable service
   right here in (Name of Town)?   Yes ☐ No ☐
   If you answered yes, please skip to question 4.

3. Is cable TV service available to you?   Yes ☐ No ☐ Don't know ☐
   To be answered only by those who are not current or previous subscribers.

   If you have answered this question, there's no need to answer any others. Just return the form to us in the enclosed envelope.

4. Why did you cancel the cable service? _____

   _____

   _____

   _____

   _____

   If you have answered this question, there's no need to answer any others.

5. What is your monthly payment to (Cable System)? $_____

6. Date cable TV was installed in your home?   _____

7. How much did you pay for the original installation? $_____

8. How many sets in your home are hooked up?   _____

9. Have you had any problem of the following kinds?
      1. Installation   Yes ☐ No ☐
      2. Quality of reception   Yes ☐ No ☐
      3. Service calls   Yes ☐ No ☐
      4. Billing   Yes ☐ No ☐
      5. Program quality   Yes ☐ No ☐
      6. Other   Yes ☐ No ☐

   If any of the "yes" boxes were checked, please explain. Use the back of this page if necessary.

   _____

   _____

   _____

   _____

10. Did you report the problem or problems to (Cable System)? Please supply details. Use the back of this page if necessary.

    _____

    _____

    _____

    _____

    _____

11. The following is a list of services (channels) now available on your cable system. Please indicate the <u>degree</u> to which you would like to continue having this service available to you. ("5" is the highest interest, "1" is the lowest.)

(List and Description of Services)	Check If You Now Receive	Low 1	2	3	4	High 5
	( )	( )	( )	( )	( )	( )

12. The following is a list of services (channels) available on cable systems <u>elsewhere</u>. Please indicate the <u>degree</u> to which you would be interested in having these available to you. ("5" is the highest interest, "1" is the lowest.)

(List and Description of Services)	Low 1	2	3	4	High 5
	( )	( )	( )	( )	( )

FIGURE 8–4. Advertising Impact Scoreboard

	✔'s	Totals
Cable		
Radio		
Television		
Outdoor		
Direct Mail		
Daily Newspaper		
Suburban Newspaper		
Word-of-Mouth		

TABLE 8–9. Arbitron Weekly Cumulative Adult Ratings

	Percent U.S. Adults Able to Receive		
	Adults 18+	Men 18+	Women 18+
CNN	57	61	53
CBN	30	26	33
ESPN	48	61	36
HBO	91	92	90
SHOW	87	89	86
SPN	35	34	35
USA	42	48	36
WTBS	68	72	64

Source: Arbitron, November 1981.

The August 1981 VideoProbeIndex, a continuing tracking study of the new electronic media, showed cumulative audience levels reported for 15 services. (See Table 8-10.)

By comparing a satellite service's 1-day cume with a 7-day cume and then with an even broader 30-day cume, an advertiser can determine what percent of the potential audience (on a monthly basis) can be delivered in a shorter period such as a week or day. For example, VideoProbeIndex reported that, in the course of 30 days, WTBS was viewed in 52 percent of its coverage area's households. It reached over 90 percent of this audience in the course of a given week—63 percent of it on a single day. (See Table 8-11.)

Many advertisers buying schedules on a given satellite network

TABLE 8–10. VideoProbeIndex Cumulative Household Ratings

	1-Day		7-Day		30-Day	
	U.S. Cable (Percent)	Coverage Area (Percent)	U.S. Cable (Percent)	Coverage Area (Percent)	U.S. Cable (Percent)	Coverage Area (Percent)
CNN	24.1	53.9	35.5	79.3	40.0	89.4
USA	6.7	16.9	13.1	33.1	16.0	40.4
ESPN	19.9	35.1	35.4	62.4	40.9	72.1
SPN	2.4	14.7	5.4	33.1	7.7	47.3
CBN	5.7	9.5	10.9	18.2	15.1	25.2
MSN	0.4	2.3	0.9	5.3	1.3	7.6
Nickelodeon	3.2	12.4	7.3	28.3	10.3	39.9
ARTS	0.9	3.5	1.8	7.0	3.9	11.6
MTV	3.6	34.6	5.8	55.7	6.9	66.2
C-SPAN	0.9	NA	2.0	NA	3.7	NA
BET	0.2	0.6	0.6	1.6	1.3	3.5
SIN	0.3	NA	0.6	NA	0.9	NA
WTBS	27.1	33.3	38.7	47.5	42.7	52.4
WGN	15.1	56.8	22.8	85.7	25.6	96.2
WOR	14.1	NA	24.4	NA	28.0	NA

Note: VideoProbeIndex developed cumulative ratings based on a total U.S. sample and projected these to each network's coverage area based upon December 31, 1981 estimates of each network's household base.

Source: VideoProbeIndex (August 1981)

want to know the *specific* cumulative audience—or reach of their *specific* buy.

At this early stage of cable research, the estimates of the four-week reach for various level schedules of cable-originated programming shown in Table 8-12 were produced. Of course, it is necessary to know the average ratings for the schedules involved. Although these data are generally not yet widely available, a good estimate is that most existing cable satellite services average a one rating or less. The exception is WTBS, which averages well over a two rating in its coverage area.

Be a Cable Research Explorer

A January 1964 statement by Arthur C. Nielsen, Sr., announced that Nielsen was terminating its network radio rating service. Mr. Nielsen said, "To my great disappointment, events beyond our control have

TABLE 8–11. WTBS Cumulative Household Ratings

	Coverage Area Percent Households
1-Day Cume	33.3
7-Day Cume	47.5
30-Day Cume	52.4

Source: VideoProbeIndex (August 1981)

TABLE 8–12. Four-Week Reach for One Cable Satellite Service (Assuming One Commercial per Program)

4-Week Target GRP Levels	Net Reach (Percent)	
	At a 1 Rating or Less	At a 2 Rating
20	8	9
25	9	11
50	14	16
75	18	20
100	21	24
125	25	28
150	28	31
175	30	33
200	32	35
225	34	38
250	35	40

changed the character of network radio. The average radio program rating has declined due to the combined effect of television and the tremendous increase in the number of radio stations competing for the available audience—approximately four-fold since 1946! Furthermore, radio listening, formerly limited largely to plug-in receivers, is now divided three ways—plug-in, battery portable, and automobile. Thus, there are now three measurement tasks instead of one!"

By changing only one or two words or key phrases, that statement could apply to cable today. It would indicate that the job of measurement is just too great to be accomplished. But to say that would be like the ostrich who sticks his head in the sand when problems occur. And cable is a place for explorers—not ostriches!

CHAPTER 9

Cable at the Local Level

"All business is local!"

This statement has long been accepted as fact by marketers of goods and services throughout the country.

For no matter how much attention is focused on the billions of dollars spent each year on national advertising, the final decision to buy or not to buy anything from a new brand of canned peas to a new car is made by an individual consumer in his or her own community.

The Local Impact of Cable

In large cities, cable will not so much provide advertisers with a new medium as it will let them deliver a unique creative treatment that existing broadcast television does not provide. This includes in-depth personal selling, advertising that's tailored to a specific environment, and affordable program sponsorships as existed in the early days of television.

In the small towns and suburban areas, however, cable will be a *brand new media form*. Television has generally not been practical either because of its waste coverage outside of these communities or because of its price tag. Cable, however, offers low-cost, informational, service-oriented advertising that is tailored to the consumers concentrated within these small geographic areas.

Each week, new towns and cities across the country are taking their first steps in the new electronic era by awarding cable franchises and beginning the process of becoming wired up. The impact of cable in each of these communities will take place at two levels—among the viewers of cable and among the advertisers who will use cable as a new selling medium. From the viewers' standpoint, a new satellite menu of news, sports, entertainment, information, and education will be further supplemented by a wide array of local community programming and neighborhood events. From an advertising standpoint, marketers will be able simultaneously to develop ways of using cable more effectively to promote and sell their products and services in these communities.

The promise of cable at the local level is to open up the video medium to advertisers in a new, affordable manner. With its dozens of channels, cable will also create program opportunities for advertisers too limited in audience appeal to be attractive to the larger broadcast television stations. The result is that local marketers will find new ways to attract the customers most interested in their products and services.

The opportunities are endless—limited only by the imagination.

What's Happening Out There?

A Quick Survey of 10 Ideas

Let's take a quick trip around the country and examine some ways in which local cable advertising has been used:

- A cable system in Moline, Illinois, leases a channel to a local real estate company. Their show is 24 hours a day of photos and descriptions of property for sale.
- In Naples, Florida, Palmer Cablevision creates local advertiser-related programs that include a garden shop, a home exchange, a money show, and a boating feature.
- In the suburbs of Boston and in New Bedford–Fall River, Massachusetts, four cable systems program 7 to 15 hours a week of local programming in which they sell advertising.
- Viacom Cablevision of Long Island solicits business advertising on its data channel. Responses were so good that they developed a special restaurant listing and car dealer listing inserts.

- Continental Cablevision of Lansing, Michigan, goes after local businesses to demonstrate their wares. A salesperson does a half hour on how to shop for a stereo. A local lumber dealer demonstrates how to build and hang a suspended ceiling.

- Albuquerque Cablevision couples local commercials run on satellite services, such as CNN and ESPN, with advertising inserts in their monthly subscriber billing statements. An advertiser's coupon or ad is inserted directly into the envelope with the subscriber's monthly statement. In other markets, cable operators are similarly finding it is quite effective to offer local clients a package combining cable spots with cable guide ads and bill stuffers.

- In Grand Rapids, Michigan, Computerland sponsored a 28-minute infomercial produced by Apple Computers. This infomercial was promoted via newspaper ads and took advantage of cable's potential to provide a detailed picture of a product sold by this computer retailer.

- CPI of Louisville carries a real estate listing and home improvement channel. During the day, there are 10-second clips with audio of homes for sale, plus longer messages for apartment and condominium complexes. In the evening, these listings are intermingled with home care programming. One of CPI's first efforts at selling advertising was its "wholesaling" of all local availabilities on the Satellite Program Network to Smith Furniture and Supply Store. Smith ran different messages to coincide with the appeal of each SPN program— women, financial, etc.—and used co-op money from General Electric for a major part of the buy. CPI promoted SPN in Louisville as the "Smith Program Network" and incorporated the store's identification into its promotion.

- In New York, Manhattan Cable offers cable classifieds to sell cameras, cars, stereos, or promote instructional or specialized services, such as acting or singing lessons, catering or counseling. They even provide a "cablegram"—a video telegram for personal messages such as birthdays, anniversaries, and graduations.

- In Chapel Hill, North Carolina, Village Cable features a *Home Shopping Channel*, which runs from 7 P.M. to midnight and offers advertisers 30 minutes for $25.00.

Turning a Live Event into a Cable Event

To utilize local cable advertising most effectively, a business must consider what the medium offers that cannot be found in any other medium. In some cases, the answer can be very obvious, yet exciting.

Here is an example. A major store in the Southwest sent out invitations to its customers to join them in a series of mini-seminars on health and beauty care. *On Monday*, there was a seminar on skin care and make-up instruction for the career woman. *On Tuesday*, the store had a program on how to update and build a wardrobe. *On Wednesday*, there was a discussion of hair care and new hairstyles. *On Thursday*, the authors of a new book on aerobics demonstrated basic exercises and dance—with a presentation of summer swimwear.

Sometimes the most obvious opportunity is the most elusive, and in this case, it was. Only after the four-day program was completed did the director of advertising realize that the promotion effort could have been extended by having the local cable system tape the different events and present them as a video promotion accompanied by the latest in summer fashions and beauty care. The store could have expanded the number of women participating in the event many times.

Lakes Cablevision—A Success Story

Lakes Cablevision, a 36-channel system that serves 9,000 households in McHenry and Lake County, Illinois (about an hour's drive north of Chicago), has taken a very aggressive stance in developing local programming and selling advertising.

Marketers have their choice of running local spots on the major satellite networks or advertising on a wider array of local programming that includes special features such as high school sporting events, band concerts, plays, local news interviews, and celebrity and political interviews.

Originating 40 hours of local programming a week, Lakes Cablevision encourages advertisers to sponsor such special features as:

- "Primary Source"—a look at the world of nutrition and foods and the people who grow and process them, with homemaking hints including recipes, decorating ideas, and crafts.
- "New and Now"—interviews with professionals in the field of beauty and personal care, focusing on hair care, fashion, nutrition, make-up, skin care, and exercise.

- "Careers"—help for the young person in choosing a career and getting a job.
- "People Issues"—a feature program that covers diverse topics of concern from alcohol and drug abuse to aging and child care.
- "Chamber Spectrum"—focusing on Chamber of Commerce activities going on around the area and featuring guests to talk about their businesses and the services they offer the community.
- "Your Chance to Live"—what to do when an emergency strikes.
- "Hook, Line and Sinker"—aimed at the fisherman with tips on when and where the fish are being caught and how to catch them.

The Selling of Local Cable Advertising

The preceding examples of advertising at the local level have been the exception rather than the rule. Today, local advertising availabilities are offered in a variety of satellite services such as:

Daytime
Entertainment and Sports Program Network
Black Entertainment Network
Cable News Networks I and II
Satellite News Channels
MTV: Music Television
CBS Cable
Cable Health Network
The Weather Channel
Satellite Program Network
USA Network

Most systems, however, are not even selling these local opportunities much less developing unique advertising concepts. Rather than invest in the sales resources and equipment necessary to succeed in the advertising business, most local systems have directed their efforts toward acquiring subscribers and improving services.

In its February 1982 issue, *Cablemarketing* reported that better than 85 percent of the systems responding to its first advertising sales survey carried satellite networks offering local advertising minutes. However, less than half of the MSO (multisystem operators) systems and only 29 percent of the independents actually attemped to sell these available spots. Furthermore, of the MSO systems that sold the local spots, only 27 percent of the minutes available each week were actually sold. And for independent systems that sold local spots, the figure was even lower. (See Table 9–1.)

Getting Started

Local systems obviously need help in evaluating the advertising potential of their markets and in setting up effective sales programs. Assistance is coming from the Cabletelevision Advertising Bureau in the form of cable advertising profiles, management reports, seminars, and consultations. Most cable operators who have been successful in selling advertising time agree that any cable system can succeed if it researches its market and hires a professional to bring in the advertisers.

It also helps to use creativity and imagination.

To introduce cable advertising in Grand Rapids, Michigan, GE Cablevision staged a one-night workshop highlighted by Las Vegas games of chance. Local advertisers played "no-loser" roulette, poker, and blackjack in which they won coupons good for free spots on the system's satellite advertising networks—CNN, ESPN, SPN and USA. The objective was for advertisers to use the coupons and, hopefully, buy additional spots. This goal was accomplished.

In another unusual kick-off of a local advertising effort, Daniels and Associates sold half of the CNN and ESPN availabilities on its Ann Arbor, Michigan, system to one local advertising agency at $1 per spot. The philosophy was that it is better to have the advertising time filled than go unsold. Furthermore, when potential local advertisers saw that their peers were using the local cable system, they were encouraged to follow suit.

When Clearview Cable in Tallahassee, Florida, began selling ads on CNN and ESPN in early 1981, they found it difficult to convince local businesses of cable's effectiveness as an advertising medium. Advertising agencies wanted documented numbers, which were not available, and Clearview even met resistance when if offered its spots on a free trial basis. To fight this problem, Clearview ran promotions, using the eight advertisers they already had managed to

TABLE 9-1. Local Selling of Spots on
            Advertising-Supported Networks

	MSO's	SSO's
Systems *Carrying* Ad-Supported Networks	87%	86%
Systems *Selling* Local Avails	48	29
Percent of Local Avails *Actually* Sold	27	25

Source: *Cablemarketing*, February 1982

sell. This got people used to seeing local ads on the two channels. Over a period of a year, this combination of plugging spots, keeping rates competitive with radio, and just building overall awareness helped turn sales around.

And, finally, there was the rather unique approach used by the small (5,000 subscriber) Westerly, Rhode Island, system of Colony Communications. Going straight to local retailers, they sold a package of 10 spots a week for 52 weeks for $3,200 to the first 15 advertisers who signed up. At an average of $6 each, the 520 spots competed with local radio. Colony's target was local radio advertisers, and they succeeded, signing up three restaurants, two auto dealers, two office supply shops, a jeweler, an appliance store, a shoe store, a florist, a home video store, a plumbing supply company, a grocery and an insurance agent.

The Local Commercial Production Problem—One Solution

A major stumbling block on the road to attracting cable advertisers who have not used television is the production of commercials. Palmer Cablevision, a cable advertising pioneer that operates systems in Florida and California, produced a package or 10 "generic" commercials for an appliance store, a jewelry store, an air conditioning/heating service, a women's and men's clothing store, a travel agency, an automobile repair shop, a furniture store, an insurance agency, and a real estate agent.

Palmer did market research to come up with the 10 different kinds of businesses known to be good broadcast advertising prospects. It then interviewed people in these businesses to find out their principal selling point. To spotlight the copy that was universally used in the various categories, they looked at the Yellow Pages.

Each 30-second spot contains 25 seconds of general audio and video content that relates to the specific business. Five seconds are reserved for the local cable system to insert the advertiser's tag with a camera card, a character generator, or a slide. This is the same concept that is found in the newspaper business with the production of local advertising mats.

Palmer markets a reel of the 10 assorted commercials for a total cost of $250 (only $25 a commercial) plus a tape charge. The advertising tapes appeal both to the smaller systems that do not have facilities to produce their own advertising messages *and* to larger systems that want to hold down advertising costs to appeal to small businesses that might not otherwise purchase cable.

Creative Associates of Louisville offers cable operators a similar service, but with a slightly different twist. They provide a reel of some 50 generic video spots free of charge to cable systems. The systems then show them to local advertisers who in turn buy the spots directly from Creative Associates, which customizes them with a "squeezoom" special effects generator or a simple voice-over. Advertisers provide copy and a camera card, and the spot is returned to them in seven days. Chuck Conrad, President of Creative Associates, says of his service: "I'd like to turn this into a giant production mail order business."

Cable Salesmanship

Unfortunately, many marketers to whom a local cable system wants to sell advertising have little actual exposure to the medium itself. A local business owner watches television, reads magazines and newspapers, listens to the radio, and sees billboards day in and day out. He may hear a lot of people talk about cable—and he may even subscribe himself—but it's still something very new and untried. This makes the job of the local system operator even tougher when it comes to selling advertising to this prospective client. The local system must recognize that it is selling an entirely new medium. How well cable works for the advertiser will depend upon the degree to which the cable system operator understands the advertiser's needs and is able to deliver what will best satisfy those needs. Nowhere is the art of good salesmanship more necessary than in the selling of cable at the local level today.

In *Cable Advertising: The First Comprehensive Guidebook for System Operators*, Texscan Corporation highlighted the six key points of the local cable sales presentation.

1. Listen to his needs. Keep him talking about his product or service. The more he expands on his needs, the better your chances of supplying the creative idea that will sell him on cable.

2. Stress the benefits. The general benefits of cable and any additional ones that are particular to your system are indispensable as selling tools. Tell him all the things cable can do for him. Educate him!

3. Use examples of what others have done. If the customer is still hesitant about using cable after you have told him what it can do for him, give him examples of successful local advertising. You might suggest that he call some present advertisers and let them do the selling for you. Word of mouth is still one of the most effective selling techniques.

4. Invite him to see your system. If the buyer has never seen cable, you have to show him what cable is. Otherwise he won't be inclined to buy.

5. Show him he's the customer. Make him feel confident that you can fill his needs with cable better than any other media could. Let him do it his way. If he has an idea for a commercial message, jump on it. That's your chance to close.

6. If you are close to a sale, you can usually tell. When he starts asking a lot of specific questions, it's time for the close. Get a commitment. Ask for the order. Try not to let him put you off until a later date. Stress the importance of getting into cable now while the industry is young and prices are low. If he insists on making the decision sometime in the future, tell him you'll call on him again. Then mention a date in a week, a month, or whatever seems appropriate. Leave something behind—a rate card, a brochure or program schedule so he doesn't forget you. (*Cable Advertising: The First Comprehensive Guidebook for Systems Operators*, p. 10.)

Local Cable Advertising and Program Production Guidelines

If cable video is to develop to its full potential at the local market level, it will be because there is established a solid working relationship between the advertiser and the local cable system. This is reflected in the following "Local Cable Advertising and Program Production

Guidelines," developed by Tom Greer, former Vice President and Creative Director of the New Media unit of J. Walter Thompson, USA, and now an independent cable producer, director, and consultant. Tom's involvement in cable extends from creating the idea to selling it and developing the execution. He recognizes that if cable advertising is to succeed, it will be because of the intertwining of efforts by the buyer and seller of the medium.

I. Identify the Local Cable System

The first step to the successful use of cable television in any community is getting to know your local cable operator. Establish what potential opportunities for advertising exist on the system. Discuss your mutual needs and interests. An ever increasing number of cable systems are providing local advertising opportunities. They generally fall into three categories:

1. Local availabilities in the national satellite, advertising-supported services. Generally, the local cable operator has several minutes of time per hour to sell in cable services such as the Cable News Network, Entertainment and Sports Programming Network, CBS Cable, The Health Channel, The Weather Channel, etc. The programs provide opportunities to advertise to a narrow target audience interested in news, sports, health, or culture.

2. Local-originated programming. Local systems generally have one or more channels that carry a variety of locally produced programs specifically targeted to needs within the community. At present these include:

Community News	Local Documentaries
Community Sports	Catalog Channels
Consumer Shopping	Interviews and Talk
Local Celebrations and Parades	Community News
Community Business Profiles	Community Cooking
	Senior Citizen Events
	"How-To"

   Advertising, as well as full-program sponsorships, is available on most of these local-origination channels. You also have an opportunity to produce programs yourself, if your message is communicated better in this manner. While satellite network programs provide targeting of audiences, locally produced programs add the ability to have a total program "brought to you by...(you the advertiser)." In addition, your local image can be enhanced by providing programming of a genuine community interest or need.

3. Automated programming and digital information. Alphanumeric channels provide information on weather, stock market quotations, community billboards, news, and sports highlights, and they usually have provisions for classified-style advertisements. While live demonstrations or programs are not possible, announcements about services, products, special sales, and so on are possible at very attractive prices. The key here is the buildup of cumulative viewing. While only a small number of viewers will watch these channels at any one time, most cable viewers will check the local weather, stocks, sports, or community information data channels at some time during the viewing week.

Some cable systems currently do not offer advertising availabilities or local-origination channels. They either have not considered the possibilities of additional revenue or currently do not have the equipment or services that allow for local advertising production. Do not give up, however! Show an interest in both the creation of local programming and in local advertising production. In the near future, most cable systems will begin to consider both local advertising and programming as they recognize it can be a significant new source of revenue. Be there first. Prepare for the future. Have your ideas ready. If you show interest now, the cable operator will knock on your door first when local advertising and program production begins.

II. Identify Local Production and Creative Talents

Most cable operators come into the industry from business and banking. While they may have the technical equipment to produce programming and commercials, they do not have the expertise in the creative and marketing areas. Look to the cable system's program director, if one exists, and also the local colleges and universities, the local newspaper and radio station, small local media shops and creative boutiques who have people who can help you create storyboards, scripts, and identify where you might begin with local creative. Local franchisees and dealers who sell goods and services for national clients will find assistance from their own national advertising agencies. Workshops conducted by the Cabletelevision Advertising Bureau, the National Cable Television Association, and local cable clubs can also be of great help. Finally, look for creative assistance from those involved in advertising production for radio and print. Remember to go to the cable operator with a good creative plan in mind. Hopefully, the cable operator will have the necessary technical plan to help you execute your predetermined concept.

III. Assess Community Needs

In order to effectively use cable TV in your community, you must first evaluate both your own and your community's needs. Based upon this evaluation, you might choose specific existing national or local program environments in which to place commercial messages, produce an informational program series involving your product or service, or underwrite a series of community cable TV events. The best use of cable TV will be different in each community, and only you and your local cable operator can really determine what specific format of message delivery is best for you.

IV. Development of Creative Concepts

If you decide to produce commercials for cable, they will look very different from what you are familiar with on broadcast TV. "Freedom of choice," based upon the large number of programs and services available to the cable viewer, dictates that commercials that don't educate, inform, or entertain the viewer probably will not be watched. Experience shows that viewers can easily switch to CNN, local weather, a sports update, stock reports, and then cut back to other programming when the commercials are over. The addition of the remote control cable converter allows the viewer to check back quickly to monitor the re-start of a selected program, virtually guaranteeing that while missing the commercial, he or she won't miss any of the show. "Zapping," the avoidance of watching commercials, is a game played seriously by those who work to see how many hours they can watch cable without seeing a commercial message.

The creation of commercials and programs for cable demands creative executions that are attractive to viewer perceptions and interests and become something that adds value to the viewer's life. Cable video can communicate in-depth the "lifestyle" benefits of your product or service that cannot be communicated in a 30-second message.

V. Media Buying, Mixing, and Promotion

Cable video is a medium like other media—TV, radio, newspapers, magazines, outdoor, penny shoppers, etc. It must be examined within the context of these other media. And when considering cable video, you should ask yourself, "What is it about my product or service that is of importance and of interest that I have not effectively been able to communicate in the past?" This is the key to cable advertising. The effective promotion of programming, particularly locally produced programs, is essential to its success. Use existing print ads and radio commercials to support your cable programs by producing short tags that tell the community about what you are offering on cable. Tie-in local sales, give-a-ways, and personal appearances with your cable ads to extend their effectiveness. Whatever you do in cable production, do it regularly, as awareness in your community about your cable programs will be cumulative, with small audiences that build each week. Cable will only be effective if you do enough, on a regular basis, to promote and tie-in your cable effort with your other media and be aware of your community's interest and needs.

VI. What Should I Pay for Cable?

Basically, you can expect to buy local cable media time at prices anywhere from $1 to $100 per advertising spot. This varies widely, and prices will also depend on whether you are talking about a 30-second announcement or a two-minute execution. If you buy a large enough schedule, many cable systems will provide production and editing at nominal charges or even for free. Negotiation with the operator of your cable system is important, and what you end up paying will be an amount that you both find acceptable. There are few rules in cable advertising and programming concerning costs, promotional trade-offs, etc. The only true rule in cable video advertising and programming is that good deals are not hard to come by. As in any media venture, potential for a long-term commitment will make a cable system operator more interested in you. If you are a long-term advertiser, you get the first opportunities for new ventures. When a cable system has programming ideas, he will approach you first, not your competitor.

Conclusion

The purpose of all advertising is to communicate a sales message that will motivate a real person to make a real purchase of a real product or service. The role of media is to communicate that message in the most efficient and effective way, whether it be the town criers who once wandered up and down our streets or any of the new media in all of their forms, present and future. The key to the successful use of the "new" media is the same as for the "old" media: offer a good product or service, identify your audience, speak to that audience in ways that are honest and meaningful, and don't be afraid of trying a new idea.

In many respects this philosophy brings our discussion of the new media full circle. As you can see, it has been virtually impossible for me to escape dealing in the old "tried and true" aphorisms, and perhaps that is most appropriate. For just as we continually seek new ways of expressing existing ideas, the new media represent powerful ways of communicating these same ideas.

In addition to a glossary and an index, the last section of this book contains three appendices. While some people think of an appendix as a place where authors "bury" what they cannot fit into an appropriate chapter of a book, that would be a mistake in this case. "The Local Cable Idea Starter Kit" will help you begin developing creative cable concepts where all good advertising communications begins—at the local level. "The Satellite Network Buying Checklist" will guide you in establishing a basic structure for evaluating cable at the national level. And the "List of New Media Networks" will give you an indication of services available today. Use them as you explore new media opportunities and experiment with new ways of letting the new media increase your marketing effectiveness.

The Local Cable
Idea Starter Kit

Earlier in this book, I noted that, in developing ideas for the effective use of cable, we are limited only by our imagination. Nowhere is this more evident than in the creation of cable concepts at the local level.

This appendix includes 104 ideas for effective cable selling concepts covering 42 product and service categories at the local level. And the challenge is the "blank space" in which you can add your own creativity.

Air Conditioning and Heating Companies

Home Insulation and Energy Conservation

_____

_____

_____

Airlines

Exercises for the Business Traveler

Great Restaurants Around the Country (World)

Getting Ready for Your Vacation

_____

_____

Appliance Stores

Household Safety Tips

Energy Saving Tips

_____

_____

_____

Art Galleries and Dealers

What to Look for in Buying Art

_____

_____

_____

Automobile Dealers

Where to Go and How to Get There

How to Shop for a Car

_____

_____

_____

Automobile Supply Stores

Do-It-Yourself Video Car Manual

_____

_____

_____

Bakeries

Entertaining for the Holidays

Tips on Party Planning

_____

_____

_____

Banks and Savings and Loans

Personal Finance

The ABC's of Getting a Loan

The ABC's of the IRA

_____

_____

_____

**Bedding
Companies**

Interpreting Your Dreams

_____

_____

_____

**Boats and
Marine
Equipment**

Boating Conditions in Your Area

Tips on Sailing

_____

_____

_____

Book Stores

Book Review of the Week

The Top 10 Books of the Week

_____

_____

_____

**Building
Material**

How to Build Almost Anything

What to Keep in Your Garage

_____

_____

_____

Burglar Alarms and Security Systems

Safeguarding Your Valuables

Protecting Your Home Against Intruders

_____

_____

_____

Clothing Stores

Community Fashion Shows

The Latest Fashions

Clothes for the Working Woman

How to Coordinate and Care for Your Wardrobe

_____

_____

_____

Cosmetics and Beauty Aids

Make-up and Beauty Hints

Taking Care of Your Face, Feet and Hands

_____

_____

_____

Dental and Medical Services

Exercises at Your Desk

Foods for a Healthier You

Taking Care of Your Mind and Body

Kick the Smoking Habit

_____

_____

Drugstores   Organizing Your Medicine Cabinet

Lists to Leave for the Babysitter

Educational   Going Back to School After 30!
Institutions

Employment   How to Interview for a Job
and Recruit-
ment Agencies   Assessing Your Strengths and Weaknesses

Writing an Effective Resume

Financial   Tips on Keeping Tax Records
Services
Understanding the New Tax Laws

Tax Deductions You May Have Overlooked

The Stock Market Report

Planning the Family Budget

Furniture   Caring for Your Furniture
Stores
Arranging Furniture in Your Home

Interior Decorating on a Budget _____

_____

_____

_____

Garden and Lawn Supplies

Planning Your Summer Garden _____

Caring for Your Garden _____

Loving Your Plants _____

New and Different Salads _____

_____

_____

_____

Grocery Stores

Holiday Food Ideas _____

Meals on a Budget _____

Recipes for the Working Woman _____

How to Fix a Last-Minute Dinner _____

Planning Your Shopping List _____

Summer Picnic Meals _____

Seasonal Food Specialties _____

_____

_____

_____

Hardware Dealers

Complete Do-It-Yourself Manual _____

How to Repair Almost Anything _____

Organizing Your Kitchen _____

Women in the Hardware Store

Health Food Stores

Eating for a Healthier Life

Putting Nutrition in Your Diet

Hobby Shops

Things to Do on a Rainy Day

Fun for the Family

Hobbies in Your Town

Home Improvements and Remodeling

Remodeling on a Budget

Redoing the Kitchen and Bath

Hotels and Motels

Spending the Weekend in Town

Entertainment Guide of the Week

Insurance Agencies

How to Buy Insurance

How Much Insurance is Enough

Luggage Stores

Packing the Most in the Least Space

Matching Your Luggage to Your Travel Needs

Movers

Getting Ready to Move

Last Minute Moving Check-List

Office Equipment

Buying a Personal Computer

Organizing Your Workspace More Efficiently

Pest Control and Exterminators

The Warning Signs of Pest Danger

When to Call the Exterminator

Photo Equipment Stores

Photography Made Easy

How to Photograph Children

Choosing the Right Camera (Film)

_____

_____

_____

Real Estate Firms

Video Home Tours

Round-the-Clock Cable Home Listings

_____

_____

_____

Restaurants

Favorite Meals of the Chef

Favorites of the Celebrities

_____

_____

_____

Service Stations

Getting Your Car Set for Winter (or Summer)

Safe Driving Tips

When Your Car Won't Start

_____

_____

_____

Sporting Goods Stores

High School Sports

Tips for (Golf, Tennis, etc.)

Sportswear Fashions

Stereos and Hi-Fi's

Video Music Show (featuring local talent)

How to Shop for a Stereo

Theaters

Entertainment Gossip News

Movie Trivia Quiz

Travel Agencies

Vacations on a Budget

Where to Go for a Weekend

Exotic Spots to Visit

Veterinarians

Taking Care of Your Dog or Cat

Taking Your Pet on Vacation

Training Your Dog or Cat

Satellite Network Buying Checklist for

(Satellite Program Network)

Description of Satellite Network Programming

Overall: _____

_____

_____

Specific Program:_____

_____

_____

Basic Programming Appeal

| Single Subject (Vertical) | _____ | or | Multi Subject (Horizontal) | _____ |
| Loyal Audience (Low Turnover) | _____ | or | Diverse Audience (High Turnover) | _____ |

Merchandising Potential

To Sales Force, Dealers, Retailers, etc.:_____

_____

_____

To Viewing Audience:_____

_____

_____

Programming Schedule

Hours a Day:_____

Days and Times:_____

_____

Program Schedule (attached)   Yes_____   No_____

Subscribers and Growth Potential

Satellite Carrier: _____

Subscribers

    Current:_____

    Year from Now:_____

    2 Years from Now: _____

    5 Years from Now: _____

Subscribers by Market (attached)   Yes_____   No_____

Subscribers by County Size (% of Total): A_____ B_____ C_____ D_____

Subscriber Characteristics (attached)   Yes_____   No_____

Program Clearance Policies

Systems Carry Full Schedule _____ Selected Programs/Segments_____

Explanation (attached) if Systems Carry

    Selected Programs/Segments   Yes_____   No_____

Advertising

Commercial Minutes Per Hour: Network_____   Local_____

Nonstandard Commercial Lengths_____

    Short Form (Under 30 seconds)_____

    Long Form (90 seconds)_____ (Two minutes or more)_____

Costs Per Commercial (30 second base)

    Rate Card_____

    Negotiated_____

Amount of Inventory Sold (estimated %)_____

Audience and Demography

Estimated Household Rating:

        Overall_____   High Estimate_____   Low Estimate_____

    Basis of Estimates: _____

    _____

    _____

Viewer Demographics (attached)   Yes_____   No_____

Maximum Schedule Reach Estimates

    1 Day_____

    7 Days_____

    30 Days_____

Network Plans for Research_____

    _____

    _____

Network Agreement to Conduct Audience Research as Part of Advertiser

Buy _____

    _____

    _____

Cable Satellite Networks

TAG/Cable Information Exchange, a research/consulting company, publishes the *Cable Advertising Directory*, providing in-depth listings of cable systems, networks, cable representatives, regional interconnects, and couponing and program guides selling advertising space.

The cable satellite network data sheets that follow appeared in the first edition of the *Cable Advertising Directory* in Spring 1982. Changes in the cable industry occur so fast and so frequently that it is necessary to update this information frequently.

Source: Katherine Connolly, TAG/Cable Information Exchange, P.O. Box 5263, New York, New York 10022

169

ARTS (Alpha Repertory Television Service)
825 Seventh Avenue
New York, N.Y. 10019
(212) 887-5000

ARTS	A full spectrum of visual and performing arts. A blend of contemporary and classical programming.
Satellite Distribution	Satcom III-R, Transponder 1. Received in 7 million homes as of 4/30/82.
Schedule	9:00 P.M. to Midnight (EST), seven nights a week.
Cost	Packages negotiated on an individual basis. Rates on request.
Projected Rating	ARTS is conducting its own audience profile studies on an ongoing basis.
TV Listings	Listed in TV Guide's 18+ (selected(regional editions which feature cable listings, as well as in many cable systems' program guides.
Launch Date	April 12, 1981.
System Fee	Free.
Advertisers	General Motors, AT&T, Polaroid, Mobil Oil, Ford Motor Co.
National Advertising Representative	John Cronopolus Director, Cable Sales 825 Seventh Avenue New York, NY 10019 (212) 887-5389 John Silvestri John Hancock Center 875 N. Michigan Avenue, Suite 1414 Chicago, IL 60611 (312) 266-1911

Black Entertainment Network
1050 31st Street NW
Washington, D.C. 20007
(202) 337-5260

Black Entertainment Network	Black-oriented programming, including films, sports, music, documentaries, women and youth.
Satellite Distribution	Satcom III-R, Transponder 9 (moves to Westar V July 1982). Received in over 9.2 million cable homes as of May 1982.
Schedule	Monday through Sunday 8 P.M. - 2 A.M. (EST).
Cost	Cost per 30-second spot: $375 average.
Projected Rating	A.C. Nielsen coincidental surveys conducted three times a year.
TV Listings	Listed in 50 editions of TV Guide; is in all major-market editions. Listed in cable program guides and newspapers. Picked up by computer listing services for distribution.
Launch Date	January 25, 1980.
System Fee	One cent per subscriber.
Advertisers	About 35 sponsors, including: Anheuser-Busch, General Electric, Westinghouse, Kellogg, Pepsi-Cola, General Foods.
National Advertising Representative	Robert Johnson President Black Entertainment Television 1050 31st St. NW Washington, DC 20007 (202) 337-5260

Cable Health Network
1211 Avenue of the Americas
New York, N.Y. 10036
(212) 719-7230

Cable Health Network	"Better living" programming offering health, science and medical features presented in an entertaining and informative manner, all designed to improve life.
Satellite Distribution	Satcom III-R, Transponder 17. Projected 4 million subscribers at launch.
Schedule	24 hours.
Cost	To be announced.
Projected Rating	To be announced.
TV Listings	To be announced.
Launch Date	June 30, 1982.
System Fee	Free.
Advertisers	To be announced.
National Advertising Representative	Robert A. Illjis VP, Director of Marketing Development and Sales Cable Health Network 1211 Avenue of the Americas New York, NY 10036 (212) 719-7376

Cable News Network
1050 Techwood Drive NW
Atlanta, Georgia 30309
(404) 898-8500

Cable News Network	Full-time programming of in-depth news reports.
Satellite Distribution	Satcom III-R, Transponder 14. Received in over 12 million cable homes as of 5/1/82.
Schedule	24 hours.
Cost	Cost per 30-second spot: $350 - $2,500 range (combination sales).
Projected Rating	A.C. Nielsen Home Video Index studies conducted in May and December, 1981; project being measured by NTI meters in June 1982.
TV Listings	Listed in many cable systems' program guides, newspapers etc.
Launch Date	June 1, 1980.
System Fee	15-20 cents per subscriber per month.
Advertisers	180+ sponsors, including: General Foods, Procter & Gamble, Merrill Lynch, Wrigley, Ralston and Quaker Oats.
National Advertising Representative	Mike Murphy VP, Sales Turner Broadcasting System 575 Lexington Avenue New York, NY 10022 (212) 935-3939
	Gary Koester VP, Advertising Sales Warner Amex Satellite Entertainment Co. 1133 Avenue of the Americas New York, NY 10036 (212) 944-5598
	David Houle 400 N. Michigan Avenue, Suite 1606 Chicago, IL 60611 (312) 661-1670
	Bill Adams 400 Renaissance Center, Suite 500 Detroit, MI 48243 (313) 259-2677
	Michael Wheeler 1800 Century Blvd. Atlanta, GA 30345 (404) 320-6808
	Doug Bornstein 90 Universal City Plaza P.O. Box 7201 Universal City, CA 91605 (213) 506-8316

Cable News Network 2
1050 Techwood Drive
Atlanta, Georgia 30309
(404) 898-8500

Cable News Network 2	News headline service with complete new cycle of reports every half-hour.
Satellite Distribution	Satcom III-R, Transponder 15. Projected 2 million homes as of 3rd quarter-1982.
Schedule	24 hours.
Cost	Cost per 30-second spot: $60 average.
Projected Rating	Special rating studies currently being scheduled.
TV Listings	Listed in many cable systems' program guides, newspapers.
Launch Date	January 1, 1982.
System Fee	0 to 5 cents per subscriber per month.
Advertisers	Currently being sold in conjunction with Cable News Network.
National Advertising Representative	Mike Murphy VP, Sales Turner Broadcasting System 575 Lexington Avenue New York, NY 10022 (212) 935-3939
	Gary Koester VP, Advertising Sales Warner Amex Satellite Entertainment Co. 1133 Avenue of the Americas New York, NY 10036 (212) 944-5598
	David Houle 400 N. Michigan Avenue, Suite 1606 Chicago, IL 60611 (312) 661-1670
	Bill Adams 400 Renaissance Center, Suite 500 Detroit, MI 48243 (313) 446-6877
	Michael Wheeler 1800 Century Blvd. Atlanta, GA 30345 (404) 320-6808
	Doug Bornstein 90 Universal City Plaza P.O. Box 7201 Universal City, CA 91605 (213) 506-8316

CBS Cable
51 West 52nd Street
New York, N.Y. 10019
(212) 975-3541

CBS Cable	Cultural programming, including performances of concerts, ballets, jazz, dramas and interviews.
Satellite Distribution	Westar IV, Transponder 3D. Received in 4 million cable homes as of 4/1/82.
Schedule	4:30 P.M. - 4:30 A.M. M-F; 5:00 P.M. - 5 A.M. S&S (EST)
Cost	Cost per 30-second spot: $550 - $1,500 range.
Projected Rating	Conducting in-depth audience measurement studies on its own, and compiling viewer profiles as well. No results yet released; research is ongoing.
TV Listings	Listed in 5 editions of TV Guide, in On Cable, and in other major cable program guides. Also carried in Sat Guide. Picked up by TV Compulog for distribution to guides and newsletters.
Launch Date	October 12, 1981.
System Fee	Free.
Advertisers	Quaker Oats, Kellogg, Exxon, Kraft, Porsche-Audi, Ford, General Motors, Shell Oil, HBO, Warner Bros., Wall Street Journal.
National Advertising Representative	Jim Joyella VP Sales 51 West 52nd Street New York, NY 10019 (212) 975-3541 Isabel Kliegman Western Regional Sales Manager Television City 7800 Beverly Blvd. Los Angeles, CA 90036 (213) 852-4000

Christian Broadcast Network
Virginia Beach, Virginia 23463
(804) 424-7777

Christian Broadcast Network	All-family entertainment including "Another Life," "National Geographic" specials, "Sing Out, America" and "$50,000 Pyramid."
Satellite Distribution	Satcom III-R, Transponder 8. Received in 16.3 million homes as of 5/1/82.
Schedule	24 hours.
Cost	Cost per 30-second spot: $650 average.
Projected Rating	A.C. Nielsen Home Video Index studies scheduled to begin on a monthly basis as of March, 1982 and will be ongoing.
TV Listings	Listed in several editions of TV Guide and in other cable program guides around the country.
Launch Date	April 4, 1977.
System Fee	Free.
Advertisers	Information not available.
National Advertising Representative	John Fernandez Director of Sales CBN National Sales 60 East 42nd Street, Suite 3910 New York, NY 10165 (212) 661-2600 Continental National Sales 360 N. Michigan Avenue, Suite 2010 Chicago, IL 60601 (312) 782-8695

Daytime
555 Fifth Avenue
New York, N.Y. 10017
(212) 661-4500

Daytime	Women's service and informational programming.
Satellite Distribution	Satcom III-R, Transponder 22. 5.6 million subscribers as of 6/15/82.
Schedule	1:00 P.M. - 5 P.M. (EST), Monday through Friday.
Cost	Packages negotiated individually. Rates on request.
Projected Rating	No formal audience measurement studies planned, but Hearst/ABC is conducting its own audience research; company also supports the NCTA/CAB audience measurement efforts.
TV Listings	Will be listed in a number of systems' program guides and TV Guide.
Launch Date	March 15, 1982.
System Fee	Free.
Advertisers	Campbell Soup Co., Kraft, Manufacturers Hanover Trust, Quaker Oats, Revlon, TWA, Sears Roebuck & Co., Nylint Corp.
National Advertising Representative	Robert Fell VP, Director of Sales 555 Fifth Avenue New York, NY 10017 (212) 661-4500 John Silvestri Regional Sales Director 875 N. Michigan Avenue Chicago, IL 60611 (312) 266-1911

Entertainment and Sports Programming Network
ESPN Plaza
Bristol, Connecticut
(203) 584-8477

Entertainment and Sports Programming Network	A sports network providing professional, collegiate and amateur sports, along with sports news, interviews and features.
Satellite Distribution	Satcom III-R, Transponder 7. Received in over 15.4 million cable homes as of 4/1/82.
Schedule	24 hours, seven days a week.
Cost	30 second spot rate: $500 - $1,600 range for 3rd quarter, 1982.
Projected Rating	Will be on A.C. Nielsen meters in Fall, 1982. Arbitron national coincidental study done November, 1981, plus other Arbitron CAMP, Nielsen phone coincidentals and other studies from Simmons, Nielsen and other third-party syndicated research organizations conducted.
TV Listings	Listed in 93 editions of TV Guide and adding more editions each week. Listed in other cable program guides and newspapers; has own program guide.
Launch Date	September 7, 1979.
System Fee	Affiliate compensation in form of co-op dollars and network compensation as of 1/1/82. Fee to systems under existing contract: 48 cents per subscriber per year.
Advertisers	200 national advertisers, including: Anheuser-Busch, Subaru, Datsun, Chevrolet, Ford, GM Trucks, Noxzema, Mennen, Gillette, Hilton Hotels, Holiday Inns, Hertz, Avis, Xerox.
National Advertising Representative	Michael O. Presbrey VP Advertising Sales 355 Lexington Avenue New York, NY 10017 (212) 661-6040 Robert McCarthy 111 E. Wacker Drive, Suite 2206 Chicago, IL 60601 (312) 938-4222 David Aubrey 8826 Dorrington Avenue Los Angeles, CA 90048 (213) 858-0516

Modern Satellite Network
1350 Avenue of the Americas
New York, N.Y. 10019
(212) 582-5500

Modern Satellite Network	Cable TV's source for consumer information programming, featuring "The Home Shopping Show," "Consumer Inquiry," and "Twice a Woman" and "Telefrance in the Morning."
Satellite Distribution	Satcom III-R, Transponder 22. Received in 4.1 million cable homes as of 3/1/82.
Schedule	10:00 A.M. - 1 P.M. (EST), Monday through Friday.
Cost	Cost per 30-second spot: $150 - $375 range. Eight-minute infomercials: $11,000 production and 5x airing.
Projected Rating	Plan either one or two A.C. Nielsen audience measurement surveys.
TV Listings	TV Compulog and Telelog pick-up listings for distribution to newspaper and guides and in most cable program guides; carried in approx. 10 editions of TV Guide and in local newspapers throughout the U.S.
Launch Date	January, 1979.
System Fee	Free.
Advertisers	About 150 national advertisers, including Scott Paper, Kraft, Swift, Wilton, Sears Roebuck & Co., General Foods, Hallmark, Pillsbury.
National Advertising Representative	Les Tolchin Director of Sales/Marketing Modern Satellite Network 1350 Avenue of the Americas New York, NY 10019 (212) 582-5500 MSN 875 N. Michigan Ave., Suite 1414 Chicago, IL 60611 (312) 951-5111 Fox, Brice & Associates 16200 Ventura Blvd., Suite 220 Encino, CA 91436 (213) 990-2950 Marketing Means & Methods 6614 Shady Brook Lane, Suite 3178 Dallas, TX 75206 (214) 368-1067

MTV: Music Television
1133 Avenue of the Americas
New York, N.Y. 10036
(212) 944-5380

MTV: Music Television	The first video music channel in stereo. Viewers see and hear artists performing visual interpretations of their music. Targeted to adults 18-34 and teens.
Satellite Distribution	Satcom III-R, Transponder 11. Received in over 4 million cable homes as of 5/1/82.
Schedule	24 hours, seven days a week.
Cost	30 second spot rate: $500 - $4,000 range. Sponsorships available.
Projected Rating	Company is conducting its own market-by-market audience studies. Plans not yet firm for national surveys.
TV Listings	None at this time.
Launch Date	August 1, 1981.
System Fee	Free.
Advertisers	100 sponsors, including: General Motors, Kellogg, MCA/Universal, Filmways, RCA Records, 7-UP, Anheuser-Busch, Jovan Fragrances, Kraft, Atari, Activision, Ford.
National Advertising Representative	Gary Koester VP Advertising Sales Warner Amex Satellite Entertainment Co. 1133 Avenue of the Americas New York, NY 10036 (212) 944-5598 David Houle 400 N. Michigan Avenue, Suite 1606 Chicago, IL 60611 (312) 661-1670 Bill Adams 400 Renaissance Center, Suite 500 Detroit, MI 48243 (313) 259-2677 Michael Wheeler 1800 Century Blvd. Atlanta, GA 30345 (404) 320-6808 Doug Bornstein 90 Universal City Plaza P.O. Box 7201 Universal City, CA 91605 (213) 506-8316

Satellite News Channel 1
41 Harbor Plaza Drive
Stamford, Connecticut 06904
(203) 964-8355

Satellite News Channel 1	Continually updated capsule summaries of regional, national and international news, with local news inserts by cable systems.
Satellite Distribution	Five transponders on Westar IV. Projected 2,000,000 cable homes as of launch date. Moves to Westar V after that satellite is launched.
Schedule	24 hours.
Cost	To be announced.
Projected Rating	None at this time.
TV Listings	To be announced.
Launch Date	June 21, 1982.
System Fee	Free if offered on basic. If offered on a tier, sliding scale applies, ranging from 0 (if tier has subscriber penetration of 75 percent) to 40 cents per subscriber per month if tier has subscriber penetration of 25 percent or less.
Advertisers	None signed yet.
National Advertising Representative	Jack Allen VP, Advertising Sales Group W Satellite Communications 41 Harbor Plaza Drive Stamford, CT 06904 (203) 964-8355 625 N. Michigan Avenue Chicago, IL 60611 (312) 951-0900 1900 Avenue of the Stars Los Angeles, CA 90067 (213) 203-0453

TAG CABLE INFORMATION EXCHANGE

Satellite Program Network
P.O. Box 45684
Tulsa, Oklahoma 74145
(918) 481-0881

Satellite Program Network	Consists of international, financial and lifestyle programming.
Satellite Distribution	Westar IV. Projected 5 million homes as of 5/1/82.
Schedule	24 hours.
Cost	Cost per 30-second spot: $100 - $500 range.
Projected Rating	Cooperating with NCTA/CAB ad hoc audience measurement committee; also studying possibilities for local measurement in conjunction with cable system affiliates.
TV Listings	Listed in several editions of TV Guide, as well as in cable systems' program guides.
Launch Date	January 1979.
System Fee	Free.
Advertisers	Approximately 100 sponsors, including: Kodak, TWA, American Airlines, Ford, Volkswagen, Sony Corp., American Express, Barrons, Wall Street Journal and Business Week.
National Advertising Representative	Norman J. Levine VP National Advertising Sales 211 East 51st St., Suite 12C New York, NY 10022 (212) 308-4120

SIN Television Network
250 Park Avenue
New York, N.Y. 10177
(212) 953-7500

SIN Television Network	Originally produced programming for the Spanish-speaking viewer. Includes variety, sports, novellas, comedies, and a national news broadcast from Washington D.C.
Satellite Distribution	Westar IV, Transponder 3x. Received in 24.7 million total cable homes including 3 million Hispanic homes, as of 6/1/82.
Schedule	24 hours, seven days a week.
Cost	Cost per 30-second spot: $500 - $5,500 range.
Projected Rating	Strategy Research Corp. conducting personal, in-home audience research every six months.
TV Listings	Listed in several editions of TV Guide plus other cable program guides.
Launch Date	September, 1976.
System Fee	SIN compensates systems 10 cents per month per Spanish-surnamed household.
Advertisers	About 300 sponsors, including: Goodyear, AT&T, Schlitz, Nabisco, Procter & Gamble, MGM Films, Marriott Hotels, McDonald's, Campbell Soup Co., Swift & Co., Pepsi-Cola, Kellogg's, Lever Bros.
National Advertising Representative	Susan Catapano Director, Affiliate Relations John Pero VP, Director of National Sales 250 Park Ave. New York, NY 10177 (212) 953-7500
	5358 Melrose Avenue Hollywood, CA 90038 (213) 463-2152
	2525 S.W. 3rd Avenue Miami, FL 33129 (305) 856-2793
	230 N. Michigan Avenue Chicago, IL 61601 (312) 782-1129
	2200 Palou Avenue San Francisco, CA 94124 (415) 641-1400
	5501 LBJ Freeway Dallas, TX 75240 (214) 980-7636
	3155 W. Big Beaver Road Detroit, MI 48084 (313) 649-5122

SuperStation WTBS
1050 Techwood Drive NW
Atlanta, Georgia 30309
(404) 892-1717

SuperStation WTBS	All-family entertainment from Atlanta, featuring movies and sports, including NCAA football.
Satellite Distribution	Satcom III-R, Transponder 6. Received in over 21.3 million cable homes as of 6/14/82.
Schedule	24 hours, seven days a week.
Cost	Cost per 30-second spot: $1,300 - $6,000 range.
Projected Rating	Monthly A.C. Nielsen national metered sample conducted on an on-going basis.
TV Listings	Listed in 83 TV Guide editions; project will be listed in all 110 editions by the end of 1982. Also listed in local newspapers and cable systems' program guides.
Launch Date	December 17, 1976.
System Fee	10 cents per subscriber.
Advertisers	About 500 advertisers, including: General Foods, Procter & Gamble, Colgate, Lever Bros., Warner-Lambert, Holiday Inns, Ford, Chevrolet, Buick, Toyota, Carter-Wallace, Wang, Xerox, Carnation.
National Advertising Representative	Gerry Hogan VP, Advertising Sales 1050 Techwood Drive NW Atlanta, GA 30318 (404) 892-1717
	Farrell Reynolds 575 Lexington Avenue New York, NY 10022 (212) 935-3939
	John Barbera 303 E. Wacker Drive Chicago, IL 60601 (312) 565-1717
	David Copp 400 Renaissance Center, Suite 500 Detroit, MI 48243 (313) 259-4622
	John Campangolo 2049 Century Park East, Suite 840 Los Angeles, CA 90067 (213) 553-9320

The Weather Channel
2840 Mt. Wilkinson Parkway
Atlanta, Georgia 30339
(404) 434-6800

The Weather Channel	Live national, regional and local weather, customized for each cable system. Localization of commercial text material provided.
Satellite Distribution	Satcom III-R, Transponder 21. Projected 3 million homes at launch.
Schedule	24 hours.
Cost	To be announced.
Projected Rating	To be announced.
TV Listings	Publishing own program guide at launch.
Launch Date	May 2, 1982.
System Fee	Free if service is offered on basic. If offered on a tier, sliding scale used with per-subscriber/per month fees ranging from 0 (if tier has 80 percent penetration) to 40 cents (if tier has 10 percent penetration).
Advertisers	None signed yet.
National Advertising Representative	Michael P. Ban VP, Advertising Sales Norman Zeller Eastern Sales Manager 767 Third Avenue New York, NY 10017 (212) 308-3055 Michael Eckert Midwest Sales Manager (312) 644-0570

A Glossary of
121 Essential New Media
Advertising Terms

Access Cable. Local channel space available either free or on a leased basis to individuals and groups in the community, including educational and governmental bodies.

Adjacency. A local or spot commercial that runs before or after rather than within the main body of a show.

Advertorial. An advertising message that is longer in length than the typical broadcast commercial message and provides in-depth information on a product, service, or company. (*See* Informercial).

Alphanumeric. Information such as news, weather, sports, stock quotations, and advertising that is projected on the screen in numerical and letter form via a character generator.

Arbitron. A research firm that gathers radio and television audience data and conducts special audience studies for cable systems and satellite networks.

Area of Cable Influence (ACI). The geographic area served by a cable company. It may include an entire town or only a portion of it in the case of most large cities.

Area of Dominant Influence (ADI). Arbitron's geographic market definition that includes all counties in which its television stations achieve their greatest audience. An ADI may include many different *Areas of Cable Influence (ACI)*.

Audience. The number or percentage of people or homes exposed to a program or other advertising vehicle.

Audience Composition. The distribution of a program's audience by age, sex, income, education, and other categories.

Audience Duplication. The number or percentage of viewers of one program who also are exposed to another program or advertising vehicle.

Audience Profile. The demographic characteristics of the people or households viewing a program, station, cable system, or other advertising vehicle.

Availability (Avails). The commercial time that a cable system, station, or network has for sale.

Average Audience (AA). A Nielsen term that reflects the number of households or persons viewing the "average minute" of a program's duration.

Basic Cable. Programming offered to a cable system's subscribers at a "basic" monthly fee.

Billboard. Usually a 3 to 10 second announcement at the beginning or end of a show that identifies the sponsor(s) of or advertiser(s) in the show.

Bonus Spot. A commercial given to an advertiser at no charge to make up for the deficiency in audience delivered by his schedule or as an incentive to get him to buy other spots.

Break. The time between or within a program used for commercials, announcements, or newsbreaks.

Broadcasting. The transmission of radio and television programs by over-the-air signals.

Cablecasting. Cable programming fed from a cable system to homes via coaxial cable rather than by over-the-air signals.

Cable Catalog. A direct response cable advertising/sales service where products are shown in catalog form on the TV screen and where viewers can make purchases via an 800 number.

Cable Penetration. The proportion of all television homes in an area that subscribe to cable. If there are 100,000 television homes in a market and 20,000 cable subscribers, the penetration level is 20 percent.

Cable Television. The transmission of television programs available from a master antenna, additional programming services distributed by satellite, and programming originated locally at a cable system to subscriber homes by cable (wires) instead of over-the-air.

Channel Capacity. The number of different channels of programming or services, some of them advertiser supported, available to subscribers of a cable system.

Character Generator. A typewriter-like device that projects information onto a television screen for the transmission of news, weather, sports, financial data, and alphanumeric advertising.

Chroma Key. A videotape effect in which one image can appear against one or more different backgrounds.

Churn. The turnover in subscribers to a cable system due to new sign-ups, cancellations, etc.

Circulation. The number of different persons or households that tune in a broadcast or cable signal during a specified period of time.

Coincidental Interview. An audience survey in which people are asked what they were viewing at the moment they were contacted.

Commercial Protection. The agreement of a network, a station, or a cable system not to schedule a competitive product or service within a certain amount of time from an advertiser.

Communications Satellite. A vehicle located 22,300 miles in space that receives and transmits a variety of network news, sports, cultural, ethnic, entertainment, and other programming to cable systems around the country.

Community Antenna Television (CATV). The early name for cable television, which referred to the distribution of television signals via a master antenna to homes that either could not receive them with their own antennaes or received them very poorly.

Composite Master. The completed videotape into which all elements have been edited.

Cost Per Point. A media planning tool that represents an estimate of the cost of one rating point of commercial time in a particular daypart, program type, or overall buy.

Cost Per 1000 (CPM). Another advertising planning and evaluation tool that represents the cost of reaching 1000 homes or individuals. It is the most commonly used method of comparing cost efficiencies for different programs or schedules.

Coverage Area. A geographic area in which a broadcast or cable signal can be received.

Crawl. A list of names that moves across a video screen. It can be used to list the dealers in a market who carry a product advertised in a cable commercial.

Crop. The desired compositon of a picture created by the camera.

Cumulative Audience (Reach). The number or percent of different people or homes reached by a schedule of programs, commercials, or advertisements over a specified period of time.

Cut. An editing technique to change one visual very quickly to another.

Dedicated Channel. A cable channel that is totally devoted to carrying a single source of programming.

Demographics. Audience composition or characteristics based on a variety of economic and social factors and used to evaluate programs.

Deregulation. Actions by the Congress and FCC to do away with or reduce restrictions involving the communications industry over which they have control.

Designated Market Area (DMA). Nielsen's geographic market definition that includes all counties in which its television stations achieve their greatest audience. Like Arbitron's ADI, it may include many different *Areas of Cable Influence.*

Diary. A research survey tool in which people record their viewing activity over a specified period of time. Either one *diary* is completed for each household or there is an individual diary completed for each viewer.

Direct Broadcast Satellite (DBS). A satellite transmission service that delivers a signal to a viewer's home directly via his *own* earth station and not through a cable system.

Direct Response Advertising. Commercials that use telephone numbers or addresses to permit viewers to order merchandise or get additional information on a product or service.

Downlink. That part of a satellite transmission system in which programming or other information is transmitted from the satellite to the ground.

Dub. To transcribe the sound or picture from one recording to another.

Earth Station. A dish-type antenna and other equipment that makes up a communications station used to send or receive programming and other information to or from a satellite.

Edit. To eliminate and join together portions of a film or tape by editing and splicing.

Electronic Newspaper. Newspaper-like textual information (news, sports, and stock listings) that may be accompanied by advertising and is transmitted on a video screen.

Fade In/Fade Out. An editing technique in which a subject appears or disappears slowly on the screen.

Fiber Optics. Thin glass fibers used to transmit light waves that, in turn, transmit information. Substantially more information may be transmitted via fiber optics than on radio waves.

Flighting. A manner of scheduling advertising so that it runs in-and-out over a period of time rather than continuously.

Fragmentation. A term that describes how the growing number of uses to which the television set can be placed (Cable, VCR, Video-disc, etc.) is resulting in a splintering of audience levels and declining broadcast television shares.

Franchise Area. The geographic area that a local government awards to a cable company and in which it provides cable service to subscribers.

Frequency. The average number of exposures received over a specified period of time by the different homes or people reached by a schedule of programs or commercials.

Gross Impressions. The sum total of the audiences of people or homes reached by each program or commercial in an advertiser's schedule of announcements.

Gross Rating Points (GRP). The same as gross impressions but expressed as the total of all rating points achieved over a period of time.

High Definition TV (HDTV). Television systems that are being experimented with and will provide sharper picture definition than the current U.S. standard, 525 lines per frame.

Homes Using Television (HUT). The total available video audience at a given time as reflected by the number or percentage of television homes watching broadcast or cable programs. (More properly called Homes Using Video)

Infomercial or **Informercial.** A long-form (2-7 minutes or more) commercial message that provides in-depth, helpful information about a product, service, or company. (Also referred to as *Advertorial*).

Interactive Cable. A two-way system in which viewers respond to what is on the screen by pushing buttons and ordering merchandise, participating in viewer polls, or requesting information.

Interconnect. Several cable systems in a given area that join to sell commercials and offer the convenience of one order-one bill advertising placement.

Jingle. A commercial message's music and lyrics.

Local Origination Programming. Programming produced by a local cable system. It usually focuses on local events and community affairs and, in many cases, offers local sponsorship opportunities.

Location. A shooting site outside of the studio.

Make Good. A commercial offered as an alternative to one that did not run as scheduled and was lost by pre-emption, withholding or mechanical failure.

Media Plan. The description of the various media that will be used to achieve an advertiser's marketing goals.

Meter. An electronic audience measurement device that automatically records broadcast or cable viewing.

Minicam. A small, portable color camera used to cover events and tape commercials outside of the studio.

Multiple System Operator (MSO). A large cable company that owns many systems throughout the country.

Multipoint Distribution System (MDS). A pay television service that transmits programming (largely movies with some specials and sports) via microwave for generally short distances to subscribers who pick them up with special antennae and converter boxes.

Narrowcasting. A cable program or program service that appeals to a "select" demographic target or special interest group rather than to a "broad" segment of the population. Examples would be "The English Channel," "Calliope" for children, and "Cinemerica" for people over 45.

Network. Broadcast stations or cable systems that are linked together by microwave, satellite, telephone lines or coaxial cable and receive programming delivered from a central distribution or production point.

Nielsen. A research firm that gathers television audience data and conducts special audience studies for cable systems and satellite networks.

Outtake. Film or taped footage that is not used in the final edited program or commercial.

Participation. A commercial that appears within the body of a program rather than between two different shows.

Pay Cable. A cable service that provides programming (largely movies, sports, and specials) for which a subscriber pays a charge in addition to his basic monthly fee.

Pay Per View Television. A pay television service for which the subscriber pays for each program viewed rather than per month.

Per Inquiry (PI). Direct response advertising in which the advertiser pays for each sale or response generated rather than for each commercial.

Persons Using Television (PUT). The total number or percentage of people watching broadcast or cable television programs at a particular time.

Pre-emption. A program that does not air but is replaced by another show or special event or a commercial that was scheduled to run but does not for any one of many reasons.

Quintile. The division of a group into equal fifths, such as cable viewing *quintiles*, from heaviest to lightest viewing *quintile*.

QUBE. Warner-Amex's two-way interactive cable service, which was developed in Columbus, Ohio, and has spread to other markets where Warner-Amex has won cable franchises.

Rating. An estimate of a broadcast or cable audience's size expressed as a percentage of all people or homes in a given demographic category.

Reach. (See **Cumulative Audience**).

Recall. A research technique in which a person is asked to remember what he saw, heard or did earlier in the day, week or longer ago.

Sample. The group of people selected to take part in a research study.

Satellite-Fed Master Antenna Television (SMATV). A minicable system for buildings connected to a private satellite antenna. It provides multichannel video to large apartment buildings and condominium complexes.

Saturation. A commercial buying technique where many announcements are run during a short period of time to reach a maximum number of homes or people.

Scatter Plan. A commercial buying technique where announcements are spread across many different programs or time periods rather than being confined to any particular one. An example would be running one commercial every four hours on ESPN over a week rather than concentrating them all in one or two football games.

Share of Audience. The percentage of all homes or people watching television at a given time who are watching a particular program, station, or channel. For example, if 28.0 percent of the homes in WTBS' coverage area have their TV sets on at 11:00 A.M. on Sunday morning and 7.0 percent are watching WTBS' Movie 17, it has a 25 percent share of audience.

Spill-in. The percentage or number of homes in one television market watching a station that originates in another market.

Spill-out. The percent or number of homes outside of a particular market watching a station located in that market. Since cable is transmitted to homes via wires rather than over-the-air, there is no spill-in or spill-out.

Splice. The combination of two or more different pieces of film into one continuous reel.

Sponsorship. The purchase by an advertiser of a sufficient amount of time in a broadcast or cable program to have product exclusivity, billboard identification, and, often, long-term price increase protection.

Spot. Commercial time available for sale by a broadcast station or cable system to local or national advertisers.

Still. A "nonmotion" photograph. Many cable systems will prepare commericals for local advertisers using stills and voice-over announcer copy.

Subscription Television (STV). A noncable pay television program service (largely movies with some sports and specials) in which a scrambled signal is sent out over-the-air and unscrambled by a decoder box on the viewer's television set.

Super. The superimposing of copy on a television screen. For example, a national automobile advertiser can provide commercial footage to a cable system and ask them to *super* the local dealer's name across the end of it.

Superstation. An independent television station (e.g., WGN in Chicago, WOR in New York, and WTBS in Atlanta) whose signal is transmitted via satellite and picked up by earth stations at cable systems across the country.

Tag. Information added locally to a commercial such as the name of a retailer where an advertiser's merchandise can be purchased. In another case, a food advertiser might take a commercial that has run on broadcast television, add a tag making available a free recipe booklet and run it in a women's oriented satellite cable programming.

Target Audience. Those persons defined in terms of their demographic characteristics or purchase behavior who are most desired by an advertiser because of their anticipated consumption of his product or service.

Teletext. The transmission of "pages" of textual information to a TV set, either over-the-air on a broadcast channel or by wire to a cable channel. With a specially adapted home television set, viewers may retrieve specific information that they desire such as news, weather, sports, finance, etc. Advertising messages and identification may accompany the transmission of the data.

Tiering. The different levels or groups of cable services available to subscribers at different monthly charges.

Total Audience. The percentage or number of homes or people who view some part of a program (usually at least six minutes). It is larger than the average audience and gives a general indication as to overall tune-in to the show.

Transponder. That part of a communications satellite which receives a signal from a specific program source and transmits it to earth stations at cable systems around the country.

Two-way Cable. The transmission of signals both ways along the cable that permits interactivity between viewer and cable system. (See **Interactive Cable).**

Universe. The total population or group being studied in a research project.

Uplink. That part of a satellite transmission system in which programming or other information is transmitted from the satellite to the ground.

Video Cassette Recorder (VCR). A machine that hooks into a television set and records programming on videotape for playback at a later date or plays pre-recorded tapes that are available to buy or rent.

Video Disc System. A machine that cannot record but that plays back pre-recorded discs available for sale.

Video Games. Electronic games that attach to the television set (and usually result in 10-year-old kids defeating 40-year-old adults by wide margins!)

Videotex. The general term for teletext and viewdata systems in which textual and graphic material is transmitted over-the-air or by cable.

Viewdata. A two-way interactive cable or telephone provided service in which data stored in a central computer is retrieved on home screens through the use of home terminals. Viewdata also permits home shopping and can offer advertising opportunities.

Viewers Per Viewing Household. The average number of people watching a program in each viewing household.

Voice-over. An announcer's voice that is heard in a commercial or program. For cable, an advertiser might take an existing broadcast commercial and add a different "voice-over" with a direct response offer.

Wide Screen. A large (up to six feet) screen television set with front or rear projection.

Wipe. The special visual effect where one object gradually replaces another across the screen.

Zoom. The camera effect whereby a subject is made to move closer in or farther away from the screen.

Sources of Additional Information

Below are listed some standard sources of information about the new electronic media. Since the number of publications and organizations devoted to aspects of the cable industry has increased manyfold during the past few years, any list will be incomplete. This list should, however, provide a good starting point.

A.C. Nielsen Company. New York.

Advertising Age. Chicago. Weekly.

Arbitron Company. New York.

Broadcasting. Washington, D.C. Weekly.

Cable Marketing. New York. Monthly.

CableAge. New York. Weekly.

CableFile. Denver. Annual.

Cabletelevision Advertising Bureau. New York.

CableVision. Denver. Weekly.

Editor & Publisher. New York. Weekly.

Electronic Media Edition, Advertising Age. Chicago. Weekly.

Marketing & Media Decisions. New York. Monthly.

Multichannel News. Denver. Weekly.
National Cable Television Association. Washington, D.C.
The New York Times. New York. Daily.
Simmons Market Research Bureau, Inc. New York.
Television Digest, Inc. Washington, D.C.
Television/Radio Age. New York. Biweekly.
TVC. Englewood, Colorado. Biweekly.
View. New York. Monthly.

Index

A

ABC research firm, 115
Access cable, 187
Ad Hoc Cable Audience Measurement Committee, 124, 127-28
Adams-Russell Company, 34, 96
Adjacency, 187
Adult entertainment broadcasts on cable stations, 46
Advertising. *See also* Direct response advertising
  advantages of, on cable stations, 51-52, 105
  CATV's potential for, 9-12
  checklist, 167
  development tips for, with cable, 88, 106
  disadvantages of, on cable stations, 52, 105
  diversity key to, in 1980s, 29-33
  effectiveness, 79
  flipping and zapping effects on, 41-43, 152
  localization of, 101
  and low-power television, 23
  message content of, 102-5
  objectives, 55, 57-58
  program for, with cable, 60-63
  response, 86-87
  revenue, on cable television, 9-10, 18
  revenue for new electronic media, 37
  selling of local cable, 145-49
  strategy, 49, 57, 60-63
  testing, 6-7
  time plentiful in cable, 45, 51, 105
Advertiser's objectives and new media, 71-99
Advertising agency
  and cable evaluation, buying, and advertising, 53-54
  cable financial strain for, 104-5
  media buys, 6
Advertising Impact Scorecard, 135, 138

Advertising Media Quiz, 50-51
Advertorial, 187
Albuquerque Cablevision, 143
Alphanumeric information, 25, 151, 187
American Association of Advertising Agencies (AAAA), 59
"American Baby" series, 78
American Home Products cable case history, 65, 68-69
Anheuser-Busch cable case history, 64, 65
Arbitron, 85, 87, 115, 116, 118, 122, 125, 135, 136, 187
Area of cable influence (ACI), 187
Area of Community Influence (ACI), 122-23
Area of dominant influence (ADI), 72, 122, 126, 188
ARTS, 73
Associated Press/CompuServe Videotex experiment, 94-95
Atari, 24
Audience, 188
  average (AA), 188
  cable, 52-53, 54, 63, 110, 115-18
  composition of, 188
  data for cable, 118, 123-24
  duplication of, 188
  measurement, 6, 83-86, 115-40, 165
  new media, 46
  profile of, 188
  ratings, 5
  research, 60, 62, 118, 120
  total, 119, 196
Availability, 188
Avails, 188
Average audience (AA), 188

B

Baird, John, 2
Basic cable, 12, 116-17, 188
Bates & Co., Ted, 85

Benton & Bowles, 12
Bicycling, 72
Billboard announcement, 188
Black Entertainment Television (BET), 73
Black out, cable requirements for, 4, 8
Bonus spot, 188
Break, 188
Broadcasting, 188
Burnett Company, Leo, 133

C

Cable Advertising Directory, 169
Cable Advertising: The First Comprehensive Guidebook for System Operators, 148-49
Cable Audience Methodology Study (CAMS), 128
Cable Coupon Network, 98-99
Cable Creative Paradox, 104
Cable Health Network, 73, 78
Cable Markets of Opportunity (CMOs), 72
Cable News Network (CNN), 53, 66, 67, 69, 73, 74, 117, 121-22, 133-34, 143, 146, 150, 152
Cable Plus, 132
Cable television. *See also* Community Antenna Television (CATV); Penetration of cable systems
  access, 187
  as advertising option, 51-52, 57-58, 74-76
  audience data for, 118, 123-24
  basic, 12, 116-17, 188
  buying time on, 54, 59, 165-67
  catalog, 88-89, 189
  with direct mail, 1, 90
  defined, 189
  as distinct from television, 39-43, 102-3
  diversity of stations with, 29-31
  fees, 18

households, 121-23
interactive, 8, 11-12, 192
as local medium, 8, 44-45, 108,
   118, 132-35, 141-54
local starter kit for, 155-64
markets, 74
measurement device for audience
   of, 125-28
needs evaluation for, 152
origins of, 2-15
pay, 12, 18, 116-17, 193
subscriber survey of, 134-35
targeting audience with, 78
tie-in promotions with, 34, 112,
   153
Cable Television Advertising
   Bureau, 72, 128
CableCash advertising experiment,
   98-99
Cablecasting, 43-44, 45, 188
Cablemarketing, 146, 147
CABLESHOP, The, 34-37, 96
Cabletelevision Advertising Bureau,
   146, 152
Cablevision, 19
CBN Satellite Network, 47, 73, 81,
   96
CBS Cable, 55, 56, 67, 73, 74, 80,
   107, 130-31, 150
Center for Research in Marketing, 6
Channel capacity, 189
Character generator, 189
Children's programming on cable
   stations, 14, 46
Chroma key, 189
Churn, 134, 189
Circulation, 189
Clearview Cable, 146
Co-axial cable, 3
Coincidental interview, 189
Commercial protection, 189
Commercials
   budgets for, 103, 104
   "generic," 147-48

growth of, on cable stations, 9
local slots for, with cable satellite
   networks, 74
modular material for, 111
multiple versions of, 107
nonstandard lengths of, 88, 89,
   91, 101
preparing, for cable, 61, 63
specialized, with cable, 102
split-60, 108-9
televised real estate, 45, 142, 143
test, 6-7
time differences in cable and
   broadcast, 45, 51, 102, 105, 110
Communications Marketing, 118
Communications satellite, 189
Community Antenna Television
   (CATV), 189
   advertising on, 5-7
   first system of, 3
   investors in, 4-5
   origins of, 2-15
Comp-U-Card, 96
Composite Master, 190
CompuServe, 94-95
Comstar satellite, 21
Connolly, Katherine, 169
Conrad, Chuck, 148
Consumer Electronics Show, 1
Continental Cablevision, 143
Copyright of broadcast programs, 5,
   8
Cost
   per advertising unit, 51
   per announcement, 101
   cable advertising, 141
   commercial production, 106, 111
   per inquiry, 89
   media, 101
   of meters for audience
      measurement, 124
   per point, 190
   per response generated, 86
   per thousand (CPM), 86, 87, 189

Coverage
area, 76-77, 121-23, 130, 138, 190
cable, in large markets, 53, 71-72
new media, table summarizing, 27
waste, 141
Cox Cable, 5, 91
CPI, 143
Crawl, 190
Creative Associates of Louisville, 148
"Creative Cable Research Kit,"
128-35
Crop, 190
Cultural events broadcasts on cable
stations, 14, 45, 130-31
Cume, 119, 138, 140
Cumulative audience, 128, 135,
138-40, 151, 153, 190
Cut, 190

D

"Daytime" satellite program service,
73, 109, 129-30
Dedicated channel, 190
Demographics, 74, 83, 190
of cable viewers, 102, 130-33
checklist of, 167
comparison of, for cable, pay
cable, and STV, 116
of "Daytime" woman, 129-30
Deregulation, 190
Designated Market Area (DMA), 72,
76, 122, 126, 190
Dial-It Telephone Polling System,
132
Diary, 123-25, 126, 128, 131, 190
Direct broadcast satellite (DBS), 30,
31, 40, 84, 191
Direct Mail Marketing Association
(DMMA), 88-89, 109
Direct Marketing Association, 95
Direct response advertising, 9, 11,
60-61, 63, 87-90, 191
used with cable, 52, 90, 94, 95,
101, 128, 131

infomercials and, 112
and toll-free numbers, 109-10
Direct video lists, 88
Dishes. *See* Earth station
Downlink, 191
Dub, 191

E

Earth station, 13, 15, 191
for direct broadcast satellite, 22
receiving, 21
Eastman CableRep, 72
*Eau Claire (Wisconsin) Leader
Telegram,* 45
Edit, 106, 107-8, 153, 191
Education and public service broad-
casts on cable stations, 14, 91
Eichoff, Alvin, 89
Electronic mail, 91
Electronic newspaper, 7, 191
ESPN cable network, 64, 66, 67, 69,
73, 74, 79, 83, 90, 109, 134,
135, 143, 146, 150

F

Fade in/fade out, 191
Family Guide to Boating Fun, 79
Farnsworth, Philo, 2
Federal Communications
Commission (FCC)
applications to, 8, 20
deregulation of cable industry by, 8
leapfrogging rules of, 8, 15
and low-power television, 22-23
pay TV and, 9
rulings of, restricting cable
expansion, 4, 5, 7
and Satcom IV space auction, 22
Fiber optics, 191
Field Enterprises, 91
Flighting, 83, 90, 191
Flipping, channel, 41, 43
Fragmentation, 4, 5, 12, 191. *See
also* Audience

Franchise area, 191
Frequency, 51, 83, 85, 101, 191

G

Galaxy I satellite, 21
GE Cablevision, 146
"Great Catalogue Guide, The,"
    88-89
Greer, Tom, 150
Gross impressions, 192
Gross rating points (GRP), 192

H

H & R Block, 25, 80-81, 112
Hallmark cable case history, 64, 65
Harvey, Bill, 110, 114, 122
Health Channel, The, 150
Hearst/ABC, 129
High Definition TV (HDTV), 192
Hijacking cable signals, 121
Historical development of new
    media, 2-15
Home Box Office (HBO), 12, 13, 14,
    21
  magazines on, 45
Home computers, 1, 24-25, 40, 96,
    114
Home security with cable, 34
Home shopping
  broadcasts for, 14, 68, 69, 96-97,
    113, 143
  services, 33-34, 90
  viewers, 131-33
Homes Using Television (HUT), 41,
    192

I

Industry Research Standards
    Committee, 128
Infomercial, 33, 60, 63, 98, 192
  creating an, 110-13
  length of, 45
  promotion of, in newspapers, 143
  wearout a problem for, 103-4

Information & Analysis, 118
IntelliVision Intelligent Television,
    24
Interactive cable, 8, 11, 12, 20,
    25-27, 192. *See also* Home
    shopping
  audience measurement with,
    123-25
  home security with, 34
Interactivity, 25-27, 33-37, 88
Interconnect, 72-74, 75, 120, 192
International Interact Corporation,
    97

J

Jerrold Corporation, 3
Jingle, 192

K

Kagan, Paul, 105
Kawasaki cable case history, 65
Kemper Insurance Company cable
    case history, 65-66
Ken-L-Ration cable case history, 65,
    66, 107
Klein, Paul L., 41
Kraft's pioneering television and
    cable advertising, 54-56, 67,
    133-34

L

Lakes Cablevision, 144-45
Lambert Advertising Agency,
    Martin, 90
Leader Tele-Cable, 45
Leapfrogging rule, 8, 15
Least Objectionable Programs, 41
"Local Cable Advertising and
    Program Production
    Guidelines," 149-53
Local market programming, 8, 44-45
Local origination programming, 11,
    150, 192

Location, 192
Low-power television, 22-23, 30
Lower Bucks Cablevision, 10

M

McDonnell-Douglas cable case
    history, 65, 67
McFarlin, Rob, 102
McLuhan, Marshall, 101
Magazines
    as advertising option with cable
        video, 51, 58
    impact of new media on, 45-48
Make good, 192
Manhattan Cable, 143
Master antennas, 3
Media. *See also* Interactivity
    appropriateness of, 144, 153
    audience approach, 82-86
    measurement services, 87
    mix, 51, 76, 87
    options, 49-69
    plan, 49-69, 99, 192
    research and evaluation, 115-40
Media Science Measurement, 118
"Medical World News" series, 78-79
Meter, 12, 123-25, 192
Metromedia, 96
Microband Corporation, 20
Microwave relay equipment, 3
Military Cable TV Network, 79-80
Minicam, 193
Modern Satellite Network (MSN),
    68, 73, 79, 96, 113, 131
Modular TV systems, 1
Monitel, 10
Movies shown on cable stations, 13,
    14, 15, 18
MRI, 115-17
MTV: Music Television, 44, 65, 68,
    73, 74, 107, 108, 133-34
Multiple system operator (MSO), 5,
    11, 146, 147, 193

Multipoint Distribution System
    (MDS), 5, 18-20, 21, 40, 84, 193
multichannel, 20-21

N

Narrowcast Marketing USA, 96, 98
Narrowcasting, 193
Nashville Network, 73
National Cable Television Associa-
    tion (NCTA), 9, 128, 152
Neighborhood TV Company, 23
Network, 193
New Electronic Media Study, 116
"New media" media values, checklist
    of, 58
News shown on cable stations, 11,
    14, 45
Newspapers
    as advertising option with cable
        video, 51, 58
    impact of new media on, 44-45
Nielsen, A. C., 7, 11, 30, 53, 72,
    84, 85, 115, 119, 121-22, 125,
    128, 131, 139-40, 193
Nielsen Coverage Areas, 76-77
Nielsen Home Video Index, 32, 117,
    118
Nielsen Television Index, 116
"Night Owl" teletext service, 91, 94

O

Ogilvy & Mather, 89
Old Spice cable case history, 65, 67,
    135
ON TV, 19
Out-of-home media
    as advertising option with cable
        video, 51, 58
    impact of new media on, 48
Outtake, 193
Over-the-air pay services. *See*
    Multipoint Distribution System
    (MDS); Subscription television
    (STV)

P

Palmer Cablevision, 142, 147-48
Parsons, Leroy Edward, 3
Participation, 193
Pay cable, 12, 18, 21, 40, 48, 83-84, 116-17, 193
Pay-per-view television, 20, 21, 193
Pay TV, 9, 12, 13, 15, 48
Penetration of cable systems
    and audience measurment, 6
    breakthrough in, 12-15
    in Canada, 42
    by mid-1970s, 12
    in 1960s, 5, 7
    by 1990, 40, 76
    table summarizing, 19, 72
    today, 17, 53
Penetration of STV and MDS, 19-20
Per inquiry (PI), 9, 193
Personal computers, 1, 24-25, 40, 98, 114
Persons Using Television (PUT), 193
Piracy with illegal cable converters, 121
PlayCable, 24
Pre-emption, 194
Premium entertainment on cable stations, 13, 14
Programming black-out, 8
Public Broadcasting System (PBS), 55
Publicity to promote cable effort, 55, 61, 63

Q

Quaker Oats Company cable case history, 65, 67-68, 81-82, 107
Questionnaire survey of cable viewers, 131-33
Quintile, 194
QUBE, 25, 194

R

Radio, impact of cable on, 43-44, 58, 147

Raisinets cable case history, 65, 68
Rating, 194
Reach, 83, 85, 128, 135-40, 151, 153, 190
    effective, 103-4
    low, 112-13
    television, 119
"Reader's Digest Do-It-Yourself Show," 46
Recall, 118, 123-25, 194
Reeves, Rosser, 77
Religious broadcasts on cable stations, 14
Rodale Press and cable series, 81

S

Sample, research, 6-7, 124, 194
Samuels, Gabe, 134
Satcom, 21-22
Satcom I, 12, 129
Satellite
    cable networks (DBS), 57, 58, 71-72, 108, 118, 123, 124, 138-40
    delivery of cable programming, 9, 12-14, 21-22, 31, 53, 142, 150
    direct broadcast, 22
    hardware, 27
    network buying checklist, 165-67
    radio transmission, 44
Satellite-fed master antenna television (SMATV), 20, 194
Satellite NewsChannels, 73
Satellite Program Network (SPN), 73, 78, 96, 143, 146
Saturation, 194
Scatter plan, 194
Scott Paper Company cable case history, 65, 68
Screen Vision, 107
Share of audience, 194
Showtime, 19
Simko, George, 12
Simmons Market Research Bureau (SMRB), 80, 115, 116, 129, 130-131

Simmons National Study of Media
and Markets, 130
Spanish Entertainment Network, 73
Special interest programming on
cable stations, 11, 14, 45, 46
Spill-in, 76, 194
Spill-out, 6, 76, 195
Splice, 195
Split cable, 6
Split-60 commercial, 108-9
Sponsorship, 195
Sports events on cable stations, 10,
13, 14, 18, 46
Spot, 195
buys, affected by distant cable
broadcasts, 6
"generic," 147-48
local, 39, 68, 76, 144
selling local cable, 145-47
Standard Metropolitan Area, 72
Stereo as home entertainment
component, 1
Sterling Manhattan Cable and
Teleprompter, 10
Still photograph, 24, 96-97, 195
Subscriber, cable, 3, 5, 9, 18, 121
Subscription television (STV), 13,
15, 18-20, 21, 31, 84, 195
demographics of, 116-17
Super, 195
Super-high-frequency (SHF) signals,
27
Super-interconnects, 74
Superstation, 8, 14, 15, 85, 195

T

TAB research firm, 115
TAG Cable Information Exchange,
72, 169
Tag for commercials, 107, 195
Target audience, 12, 23, 51, 195
with cable, 78, 101
demography, 63
mass-market, 77-78, 110

non-mass-market, 78-82, 110
TeleShop'r service, 10
Teletext, 25, 27, 196
Televised Real Estate, 45
Television, broadcast
advertiser expenditures for, 104,
120
as advertising option with cable
video, 51, 58
advertising treatment differences
for cable and, 141
development of, 2
hardware changes in, 27
impractical for small-town
advertisers, 76, 142
low-power, 22-23
in 1980s, 17
production schedule for
commercials on, 107
viewer contact more necessary
than, with cable viewers, 102,
103
viewing levels, 40-41
Television Digest, 19
Thompson, J. Walter, 11, 34, 96,
102, 108, 134, 150
Tie-in promotions, 34, 52, 112, 153
Tiering, 18, 196
Time shifts, 31, 40, 57
Toll-free (800) numbers, 33-34, 36,
89, 96, 109-10, 131
Total audience, 196
Transponder, 196
Turner, Ted, 8, 15, 44
Tush, Bill, 132
Twentieth Century-Fox cable case
history, 65, 69
Two-way cable. *See* Interactive
cable

U

UHF stations, 5, 18
Universe, 196
Uplink, 196

Upscale viewers, 41, 46-47, 52, 80, 102, 130-32
USA Cable Network, 64, 66, 69, 73, 81, 107, 134, 146
UTV Cable Network, 35-37, 73, 96

V

Valentino, Thomas J., 108
Viacom Cablevision, 127, 142
Video cassette, 31
    advertising on, 114
    annual report on, 98
    cable strategy of, 57-58
    as new medium, 23
    news release on, 97
    real estate listings on, 97
    recorder (VCR), 1, 23, 46, 196
    three-dimension, 27
    timeshifts of viewing patterns with, 32-33
    use of, by 1990, 40
Video disc system, 1, 31, 196
    advantages of, over VCR's, 23-24
    advertising on, 114
    audience for, 46, 48
    cable strategy with, 57-58
    development of, 2
    trade show use of, 98
    use of, by 1990, 40
Video games, 1, 24, 31, 40, 96, 196
Video newspaper. *See* Electronic newspaper
Video Shopper, 33-34
Video Stock Shots, 108
Videography, 92-93
VideoProbeIndex, 116, 118, 138-40
Videopublishing, 46

Videotex, 25, 40, 82, 91-95, 114, 196
Viewdata, 25-27, 197
Viewer choices, 29-31
Viewers per viewing household, 197
Viewing patterns, television
    new media changing, 41-43
    1980s changing, 31-33
    video recorder changing, 40
Viewtron, 25, 26, 27
Village Cable, 143
Voice-over, 197

W

Wanamaker, John, 78
Wearout, viewer, 103-4, 111, 112
Weather Channel, The, 73, 81-82, 150
Weinstein, Carl, 72
Western Union Westar satellite, 21
WGN (Chicago), 73
Wide screen, 197
Wilton Enterprises cable case history, 65, 69
Wipe effect, 197
"Woman's Day USA," 81, 82
Women's entertainment on cable stations, 14, 46, 78, 81, 129-30
WOR (New York), 73
WTBS, superstation, 8, 15, 34, 44, 53, 67, 69, 73, 85, 116, 132, 133, 138-40

Z

Zapping, 42-43, 152
Zoom effect, 197
Zworykin, Vladimir, 2